The
South Atlantic
Quarterly

SAQ

T0261567

120:1 · January 2021

Solarity
Darin Barney and Imre Szeman, Special Issue Editors

Imre Szeman and Darin Barney

Introduction: From Solar to Solarity

Solar energy is a promise. It has come to be seen as a solution to the present—perhaps the ultimate one, the only one. It names a future that we are already (all too slowly) creeping into, one which seems to have none of the problems that trouble or worry the present. In the face of the apparently intractable, unsolvable challenge of global warming, solar stands ready to ride to the rescue, shattering the tragic link between energy (fossil fuels) and CO_2 production. Thanks to solar, energy use will no longer be accompanied by doubt and anxieties about global footprints or melting Arctic ice. Solar energy announces the end of one historical cycle and the opening up of another. Against the looming eschatology announced by virtually every report about the juggernaut of global warming, solar reignites progress and offers it to the people and places left behind by fossil-fueled modernity. If that wasn't enough, there is at least the hint that solar might also solve all manner of political and social divisions, creating a world awash in energy, justice, and life.

As this chain of promises might suggest, solar energy is also emerging as one of the sharpest and most powerful of ideologies, blurring concept, fantasy, and infrastructure together in a

The South Atlantic Quarterly 120:1, January 2021
DOI 10.1215/00382876-8795656   © 2021 Duke University Press

manner that makes it difficult to disentangle solar fiction from solar reality. The career of solar as a powerful ideological form has an extended history. Two decades ago, one of the chief proponents of solar, Hermann Scheer, spoke repeatedly of its use as the obvious rational choice for human energy needs, which is why it puzzled him that polities simply didn't choose to immediately make the switch. "Although technology is normally thought capable of anything," Scheer writes, "it remains for most people inconceivable that it might achieve the relatively simple task of meeting energy needs from the sun" (1999 [2002]: 63). Even earlier, scenario planner and futurist Herman Kahn (1976: 83) noted that because of solar "the world need not worry about energy shortages or costs in the future. And energy abundance is probably the world's best insurance that the entire human population (even fifteen to twenty billion) can be well cared for, at least physically, during many centuries to come." Today, once again, repeated confirmation of solar's possibility—its growing efficiencies and falling costs—is imagined as being enough to bring about its reality.[1] The message that solar is waiting in the wings to save us is a powerful message that cuts across social and class divisions, the public and the private, and traverses political and national borders alike. The belief that the sun can meet our energy needs *and* take care of humanity, and can accomplish both via technology alone, is at the heart of the promise of solar. It is a belief that foregrounds the sun's abundance and the human ingenuity required to harness it, while relegating to the shadows the very real challenges of producing energy transition on a global scale. As Nicole Starosielski observes in her essay included here, in turning toward the sun we tend to turn away from the ground and the body, where the actual work of energy transition gets done.

This is not to say that the promise of solar is entirely a fiction. Among the real-world facts about solar is that people in many locations have seized upon it as a means to generate energy prosperity and autonomy where before there was poverty and dependency. The vision statement of Soulardarity, a community organization in Highland Park, Michigan, testifies to this:

> The neighborhoods long afflicted by blight and neglect are coming back. Vacant lots are now parks, gardens, new affordable housing, and solar farms that generate power for the city and its surrounding communities. The solar-powered streetlights installed in 2017 continue to shine. The recent integration of wifi has allowed every resident of Highland Park to access affordable internet service. Civic participation is at record heights and climbing. (Soulardarity)

This is no idle boast. As Shane Brennan (2017: 179) reports, Soulardarity's *Let There Be Light* campaign to replace infrastructure repossessed by the local power authority with a network of neighborhood-owned, democratically-controlled, solar-powered streetlights goes beyond illuminating the streets: "The Soulardarity streetlights are a way to begin [a] larger conversation about how to move toward infrastructural self-determination at the community level, and thereby upend dominant racial and economic power structures that have placed many residents in the precarious position of not being able to access or afford essential utilities." And Soulardarity is not alone, as affordable, roof-top domestic solar, under a variety of infrastructural and economic models, is surging in racialized and low-income neighborhoods across the United States (see Jackson 2017). In Canada, the transition to community-owned and operated solar-powered electricity generation is being led by Indigenous and remote communities for whom these projects represent concrete steps in the direction of energy decolonization (Kinder, this issue; Rezaei and Dowlatabadi 2016). As for rapid transition to large-scale solar power as a coordinated response to pollution, climate change and energy poverty, it is the Global South—in particular, India—that is the leading the way (Buckley 2019). It would be a cruel—and equally ideological—response to the promise of solar to dismiss the transformative material and political potential of these developments.

This special issue of *SAQ* on "Solarity" interrogates the current ways in which solar is understood and the multiple uses to which it is being put in a period of energy interregnum. The essays collected here range widely in theme and theoretical approach, in part because they constitute some of the first articulations of solar within critical theory and cultural studies, broadly understood.[2] These essays do more than deflate or uncover the premises and fantasies contained within the complicated concept-object named "solar." They also build upon the social and political openings that the transition to solar—and to renewable energy more generally—will inevitably generate. One of the premises of this exploration of the emergent politics of solar is that energy plays a foundational role in the constitution of cultural, social, and political possibilities. This claim should not be taken as akin to an energy determinism. Rather, it constitutes an important and essential corrective to the failure (until recently) to account for the impact of energy on the organization of everything from lived experience to geopolitical decision-making. This insistence on the foundational significance of energy is one of the guiding principles of research in the energy humanities (Szeman and Petrocultures 2016). But the strong claim for energy goes well beyond this field and is now an accepted principle of virtually all forms of energy studies. In their reflections on the first five years of the journal *Energy Research and Social Science*, the editors assert this principle strongly and without qualification. "We cannot fully account for any aspect of socio-political organization without understanding the crucial role played by energy on a substantive level," they write, "specifically, the shift from coal to oil as the most globally influential fuel, and latterly from oil to what emerges from our current energy transition" (Van Veelen 2019: 2).[3] With the insights that have been generated through the study of the first transition, "Solarity" begins an investigation of this second transition, which has yet to be adequately explored and critiqued.

The future of energy has in many respects already been conceded to renewables, with solar leading the charge. What remains to be understood about solar is not the amount of energy it can produce, or whether it is truly an adequate replacement for fossil fuels, but the conditions of social and political possibility solar might generate, and the relationships, strategies and conflicts that will attend this latest and perhaps last energy transition. In his essay included here, Dominic Boyer points out solarity is, or ought to be, agnostic as to the social forms it takes. The concept of solarity developed here refers to a social condition, not an energy source. The point of thinking about solarity is to consider the transition to renewables as a process involving political and economic structures and relationships, as well as social and cultural upheaval. The transition to solar might well constitute a momentous opportunity for left politics—a genuine opening produced by a transition away from a fossil fuel-powered modernity, including the form of liberal democratic capitalism that has imagined itself to be synonymous with freedom, but which in fact has been predicated on extraction, environmental destitution, inequality, exclusion and violence against the human and non-human inhabitants of this planet. But it also might simply reproduce a variant of the same system with a new form of energy that would confirm the ideologies of progress and techno-utopianism so important to liberal capitalism. Solar might avert the end of the world. But a solar future might well come into being in ways that strengthen the present's grip on the end of history.

Scheer and others imagine that the advent of solar will undo existing geopolitics in a flash, producing a sequence of distinct localities, each of which can produce their own energy and determine their own needs and desires.[4] The missing piece of the puzzle—a big piece—is the struggle, already emerging, over who controls solar, makes decisions about it, and to what ends. Whether it is treated as evidence of the environmental legitimacy of capitalism, or produces political openings onto new forms and modes of solidarity, the transition to solar is unlikely to be smooth, untroubled by the weight of history or by extant geopolitics. From the perspective of those who invest solar with the hope of a substantially *different* future, it is not clear that a *smooth* transition is desirable. Energy transition has the potential to disrupt existing sites of power and influence within industry and politics that have developed in conjunction with fossil fuels; those companies and governments whose ambitions have been fueled by coal and oil are unlikely to give up their positions of power easily or at all (Goldthau et al. 2019). Even as some states within the US (and indeed, the US military) have initiated renewable energy projects in order to shift away from fossil fuels, the US federal government offers a prime example of a fossil fuel regime desperately trying to hold onto its current configuration of power at a moment when the US is once again the world's largest producer of oil.[5] In Canada, transition has been figured differently, but to the same effect, with a proposal to use a proportion of revenues from continued oil extraction to fund renewable energy projects, thus satisfying both those who want extraction to continue and those who don't. In the wake of the coronavirus crisis, these plans seem to have dropped away, with the federal government renewing its commitment to the Canadian oil industry via an enormous new infusion of financial resources (Fife, Graney, and Cryderman 2021).

The inevitable push back of oil regimes will constitute a key site of political struggle in relation to solar, but it would be a mistake to think this wholly defines the political horizon, as if all we have to do is convince large-scale state and commercial actors that energy transition is in their long-term interest. Achieving this would not guarantee power shifts of the political sort, nor the outcomes we associate with social and environmental justice. As mentioned above, India is a world leader in large-scale transition to solar; its current government (whose image and fortunes have been bolstered by its audacious solar infrastructure program) is also among the world's most regressive (see articles in the Against the Day section of this issue). Critics of the proposed US Green New Deal (variants of which exist in other countries) have pointed out that a solar transition might actually save existing modes and formations of power rather than imperiling them. Thea Riofrancos has argued that the policy framework currently contained in the US Green New Deal

> will amount to a corporate welfare windfall of investment opportunities lubricated with tax breaks and subsidies; public-private partnerships; infrastructure outlays that will stimulate real estate development; and, a jobs

guarantee that will stimulate consumption—a win-win for the state and capital, but, by leaving the underlying, growth-addicted, model of accumulation untouched a loss for the planet and the communities most vulnerable to climate crisis and eco-apartheid (2019: n.p.).

As Jamie Cross's essay in this issue on solar initiatives in West Africa shows, although there are some states and some industries that want to keep things the way they are, there are others who see an advantage in supporting the creation of a new sector—renewable energy—that promises an enormous return on investment to those who can get there first (see Hirst 2020). For clean tech investors, the environmental benefits of solar constitute a wonderful marketing tool. As organic is to food, solar is to energy—a virtuous substitute that dispels fears and anxieties about who is doing what with solar and why.

These are the stakes of solarity, and they are what make energy transition a political, not just a technical, field. There are and will be attempts to limit or slow down a transition on the part of those committed to fossil fuels. There are and will be other attempts to take political and economic advantage of energy transition. And there are and will be still others that recognize this moment of transition as one in which possibilities outside of the vicious practices of capital and the neoliberal state might both be awakened and sustained at the site of energy and infrastructures. Timothy Mitchell's (2011) influential account of "carbon democracy" has established that energy materials and infrastructures are important media of political organization and contestation. This is doubly so under conditions of environmental duress and energy transition.

The transition from oil to solar that is already underway, in conjunction with the imperatives of climate change, has reawakened attention to the consequences of energy extraction and transportation for in-line communities and the environment. The attempt to expand oil infrastructures across borders, through communities, and into Indigenous territories in Canada, the United States, and elsewhere on the globe has generated some of today's most powerful examples of active resistance against the presumptions of extractive capitalism and environment destitution (see Estes 2019). Infrastructures of renewable energy, too, have been sites of political formation and contestation (Barney 2019; Howe and Boyer 2019; Swyngedouw 2015). The same will undoubtedly be true of solar energy infrastructures, across a spectrum ranging from opposition to innovation: solar infrastructures will be media for preventing social, political and material change, and media for accomplishing it.

If these varied sites of contestation sound like they might generate an impasse, it is with good reason. "Impasse is a situation of radical indeterminacy where existing assumptions and material relations can no longer hold or sustain us," the Petrocultures Research Group writes in *After Oil* (2016: 16). Interestingly, the uncertain terrain of energy transition overlays an institutional terrain in which established forms of political representation and legitimacy also appear to be exhausting themselves, leading in many parts of the world to the resurgence of anti-democratic populisms premised on nativism. Yet, as Simpson and Szeman argue in their essay in this issue, it is possible this compounding impasse also provides conditions for other forms of politics to emerge in tandem with an infrastructural transition whose implications are not yet fully mapped. Solarity insists on an understanding of energy as more than the fuel that powers the engine of society, but also as a force in the destitution and constitution of social and political forms.

The solarity we envision is committed to the core impulse guiding left politics, which is the struggle for equality and social justice against the rapacious force of extractive capitalism. The realities of environmental racism and the implication of energy extraction in ongoing colonial histories mean that any concept of solidarity worth the name must begin from the experiences of those whose bodies and relations have been made expendable through the brutality of extraction, and who stand to suffer most greatly from the accelerating climate and environmental effects of fossil fuels (Yusoff 2019). This means that solarity begins in solidarity with Black and Indigenous people in the Americas and elsewhere, with racialized and impoverished communities in the so-called Global South, with women, with care-workers, with those who have been disabled by their environments, and with the non-human others previously relegated to the exploitable domains of mere objecthood (Cross 2019b; Ray 2017; Wilson 2018; Whyte 2017). The first imperative of solidarity in relation to these will be to stand aside and accept their leadership in the struggle against the global fossil fuel regime, and in the development of radically alternative practices, relations, and infrastructures of solarity. This might include putting our (in *our* case: white, male, affluent) bodies and our accustomed ways of living on the line, as others have done for so long with theirs. As Nandita Badami argues in her provocative essay in this issue, we may need to turn from Eurocentric ideas about the sun and "enlightenment" to a solarity of endarkenment. The second imperative is to think and work together to develop political and economic forms that facilitate, nurture, and manage egalitarian solarities, as an energetic base for even more widespread social transformation. A solarity animated by solidarity will require humility, patience, and courage, especially on the part of those for whom petrocapitalism has delivered mostly comfort, convenience and impunity. This, and not just our fuel source, has to change. What we understand as solarity has been voiced by our colleague, Warren Cariou. His call for an "indigenized philosophy of energy" shaped in relation to an ethics of "kinship, respect, and responsibility" (2017: 18–19) speaks to how we imagine the solidarity of solarity.

The coming age of solar has the potential to redefine many of the limits of the present. Christophe Bonneuil and Jean-Baptiste Fressoz note that "the suburbanization and motorization of Western societies are certainly the most massive example of a technological and civilizational choice that is profoundly suboptimal and harmful" (2016: 113). Fossil fuels were key to the fracturing of social life over the course of the twentieth century, due to their role in making it possible to abstract experience from space; in the process, this form of energy made social subjects into private individuals, a shift at once structural and phenomenological, and an important development for the political project of neoliberalism, as both ideology and reality. Solar energy may well play an essential role in undoing the harmful "civilization choice" that has placed the planet into a climate crisis that it can only hope to adapt to and mitigate.

The new communities of experience made possible by solar have been imagined in multiple ways, from off-grid, individualized energy pods occupied by libertarians to solar anarchism shaped around a radical collectivism (animated, perhaps, by Peter Kropotkin's conception of "mutual aid" (1902 [1972])), and from localized nodes of same-old, same-old capitalism to fully-fledged global communism, with solar playing the role of those technologies Karl Marx saw as important for true collective freedom. All these visions of post-solar societies forget what the philosophers Teré Vaden and Antti Salminen have pointed out in their work. "De-fossilized subjects do not see themselves independent from larger natural and social wholes," they write, "up to the point that the term 'subject' may not apply to them, at least from a modern perspective" (2018: 46–47). If struggles against liberal capitalism and colonial extraction are successful in producing a real transition, a genuine shift to solar, they will re-constitute the subject of modernity in a fundamental way. As several of the essays included here suggest—particularly those by Joel Auerbach, Amanda Boetzkes, and Eva-Lynn Jagoe—we do not yet have the language through which to understand the solarities to come. The interventions collected here are an attempt to learn a new language for a community that will only start speaking it when fossil fuels are abandoned or the oil wells run dry.

Notes

Several of the authors included here took part in the event *Solarity: After Oil School II*, held in May 2019 at the Canadian Centre for Architecture in Montréal, Québec, which was organized by the Petrocultures Research Group with the support of the Social Sciences and Humanities Research Council of Canada. We are grateful to participants in this event for discussions that shaped these essays. We are also grateful to Kristi Kouchakji for editorial assistance in preparing this special issue.

1   The examples are legion, but see Mark Z. Jacobson and Mark. A. Delucchi (2011). The Leap Manifesto (2015), a document intended to rally Canadians to the cause of energy transition, cites the work of Jacobson and Delucchi as its primary source of evidence for the possibility of solar to fuel the world, and to do so effortlessly.

2   The critical investigation of solar has benefited from the work of a number of the contributors to this issue, including Daniel Barber, Amanda Boetzkes, Jamie Cross, and Gökçe Gunel. See Barber 2016; Boetzkes 2017; Cross 2019a, 2019b; and Gunel 2019. See also Cross, Mulvaney, and Brown 2020; and Mulvaney 2018.

3   Speaking with respect to the specific import of Timothy Mitchell's *Carbon Democracy* (2011) to energy theory, the editors write: "we cannot fully account for the production and consumption of oil without reckoning with the strategic, political, economic and social forces implicated by, and implicated in, the transition from coal to oil" (Van Veelen 2019: 2). Mitchell offers a rich and influential account of the previous energy transition. An examination of solar demands that we look forward to the next transition with the same points in mind: the ranges of forces implicated by, and implicated in, the transition from oil to solar.

4   Dominic Boyer (2018: 182) largely affirms Scheer's views in "Revolutionary Infrastructure," pointing to Scheer's views as one of the potential "effective antidotes to the paralytic agents of carbon epistemics" and as a prototype for "revolutionary trajectories that are not predicated on the growth and motor ideologies of the fossil fuel era." Andreas Malm (2016: 374), however, remains unconvinced. "If Scheer is correct about the ramifications of a transition to the flow, it stands in antagonistic contradiction to the logic of global capital, for *the means of production would have to be shackled to communities formed around energy nuclei.* The formula that once brought steam to ascendancy would have to be inverted. Capital would need to carry the people to the power, rather than placing the power amongst the people as it has been doing for the past two centuries, and never with greater vigour than in the current stage of abstract space."

5   A report by the Global Gas and Oil Network (2019) indicates that over the next five years (2020–24), oil and gas companies are planning to invest $1.4 trillion USD in new extraction projects. Eighty-five percent of this money will be spent by North American companies (primarily Canada and the US). This planned infrastructural spending has been upended by the coronavirus crisis, at least for the time being.

References

Barber, Daniel. 2016. *A House in the Sun: Modern Architecture and Solar Energy in the Cold War.* Oxford: Oxford University Press.

Barney, Darin. 2019. "Beyond Carbon Democracy: Energy, Infrastructure, and Sabotage." In *Energy Culture: Art and Theory on Oil and Beyond*, edited by Imre Szeman and Jeff Diamanti, 214–28. Morgantown, WV: West Virginia University Press.

Boetzkes, Amanda. 2017. "Solar." In *Fueling Culture: 101 Words for Energy and Environment*, edited by Imre Szeman, Jennifer Wenzel, and Patricia Yaeger, 314–17. New York, NY: Fordham University Press.

Bonneuil, Christophe, and Jean-Baptiste Fressoz. 2016. *The Shock of the Anthropocene: The Earth, History, and Us*, translated by David Fernbach. New York: Verso.

Boyer, Dominic. 2018. "Revolutionary Infrastructure." In *The Promise of Infrastructure*, edited by Nikhil Anand, Akhil Gupta, and Hannah Appel. Durham, NC: Duke University Press: 174–86.

Brennan, Shane. 2017. "Visionary Infrastructure: Community Solar Streetlights in Highland Park." *Journal of Visual Culture* 16, no. 2: 167–89.

Buckley. Tim. 2019 "Good news for climate change: India gets out of coal and into renewable energy," *Bulletin of the Atomic Scientists*, December 16. thebulletin.org/2019/12/good-news-for-climate-change-india-gets-out-of-coal-and-into-renewable-energy/.

Cariou, Warren. 2017. "Aboriginal." In *Fueling Culture: 101 Words for Energy and Environment*, edited by Imre Szeman, Jennifer Wenzel, and Patricia Yaeger, 17–20. New York: Fordham University Press.

Cross, Jamie. 2019a. "Selling with Prejudice: Social Enterprise and Caste at the Bottom of the Pyramid in India." *Ethnos* 84, no. 3: 458–479.

Cross, Jamie. 2019b. "The Solar Good: Energy Ethics in Poor Markets." *Journal of the Royal Anthropological Institute* 25, no. S1: 47–66.

Cross, Jamie, Dustin Mulvaney, and Benjamin Brown. 2020. *Capitalizing on the Sun: Critical Perspectives on the Global Solar Economy*. Baltimore: Johns Hopkins University Press.

Estes, Nick. 2019. *Our History Is the Future: Standing Rock Versus the Dakota Access Pipeline, and the Long Tradition of Indigenous Resistance*. New York: Verso.

Fife, Robert, Emma Graney, and Kelly Cryderman. 2021. "Ottawa Prepares Multibillion-Dollar Bailout of Oil and Gas Sector." *Globe and Mail*, March 19. theglobeandmail.com/politics/article-ottawa-prepares-multibillion-dollar-bailout-of-oil-and-gas-sector/.

Global Gas and Oil Network. 2019. *Oil, Gas, and the Climate: An Analysis of Oil and Gas Industry Plans for Expansion and Compatibility with Global Emission Limits*. ggon.org/wp-content/uploads/2019/12/GGON19.OilGasClimate.EnglishFinal.pdf.

Goldthau, Andrea, et al. 2019. "How the energy transition with reshape geopolitics." *Nature* 569, May 2: 29–31.

Gunel, Gökçe. 2019. *Spaceship in the Desert: Energy, Climate Change, and Urban Design in Abu Dhabi*. Durham, NC: Duke University Press.

Hirst, Laura. 2020. "Oil Slump May No Longer Be a Curse for Renewable Energy." *Bloomberg Green*. April 24. bloomberg.com/news/articles/2020-04-24/oil-slump-may-no-longer-be-a-curse-for-renewable-energy?sref=xpkx9tNh.

Howe, Cymene, and Dominic Boyer. 2019. *Wind and Power in the Anthropocene*. Durham, NC: Duke University Press.

Jackson, Derrick Z. 2017. "Affordable Solar Power Is Coming to low-Income, Minority Neighborhoods." *Undefeated*, April 21. theundefeated.com.

Jacobson, Mark Z., and Mark A. Delucchi. 2011. "Providing All Global Energy with Wind, Water, and Solar Power, Part I: Technologies, Energy Resources, quantities, and Areas of Infrastructure, and Materials." *Energy Policy* 39: 1154–69.

Kahn, Herman, William Brown, and Leon Martel. 1976. *The Next 200 Years: A Scenario for America and the World*. New York: William Morrow and Co., Inc.

Kropotkin, Peter. 1902 (1972). *Mutual Aid: A Factor of Evolution*, edited by Paul Avrich. London: Allen Lane.

Leap, The. 2015. "The Leap Manifesto." leapmanifesto.org.

Malm, Andreas. 2016. *Fossil Capital: The Rise of Steam Power and the Roots of Global Warming*. New York: Verso.

Mitchell, Timothy. 2011. *Carbon Democracy: Political Power in the Age of Oil*. New York: Verso.

Mulvaney, Dustin. 2018. *Solar Power: Innovation, Sustainability, and Environmental Justice*. Berkeley: University of California Press.

Ray, Sarah Jaquette, et al. eds. 2017. *Disability Studies and the Environmental Humanities: Toward an Eco-Crip Theory*. Lincoln: University of Nebraska Press.

Rezaei, Maryam, and Hadi Dowlatabadi. 2016. "Off-grid: community energy and the pursuit of self-sufficiency in British Columbia's remote and First Nations communities." *Local Environment* 21, no. 7: 789–807.

Riofrancos, Thea. 2019. "Plan, Mood, Battlefield: Reflections on the Green New Deal." *Viewpoint Magazine*, May 16. viewpointmag.com/2019/05/16/plan-mood-battlefield-reflections-on-the-green-new-deal/.

Scheer, Hermann. (1999) 2002. *The Solar Economy: Renewable Energy for a Sustainable Global Future*. London: Earthscan.

Soulardarity. 2016. *Let There Be Light: Building a Brighter Future in Highland Park*. Highland Park, MI: Souludarity. soulardarity.nationbuilder.com.

Swyngedouw. Erik. 2015. *Liquid Power: Contested Hydro-Modernities in Twentieth-Century Spain*. Cambridge, MA: MIT.

Szeman, Imre, and the Petrocultures Research Group. 2016. *After Oil*. Morgantown: West Virginia University Press.

Vaden, Teré, and Antti Salminen. 2018. "Ethics, Nafthism, and the Fossil Subject." *Relations* 6, no. 1: 33–48.

Van Veelen, Bregje et al. 2019. "What Can Energy Research Bring to Social Science? Reflections on Five Years of *Energy Research and Social Science* and Beyond." *Energy Research and Social Science* 57: 1–6.

Whyte, Kyle. 2017. "Indigenous Climate Change Studies: Indigenizing Futures, Decolonizing the Anthropocene." *English Language Notes* 55, no. 1–2: 153–62.

Wilson, Sheena. 2018. "Energy Imaginaries: Feminist and Decolonial Futures." In *Materialism and the Critique of Energy*, edited by Brent Bellamy and Jeff Diamanti, 377–411. Chicago: MCM.

Yusoff, Kathryn. 2019. *A Billion Black Anthropocenes or None*. Minneapolis: University of Minnesota Press.

Nicole Starosielski

Beyond the Sun:
Embedded Solarities and Agricultural Practice

Solar theory often invokes a subject basking in the sunlight, arms stretched out, and face to the sky. Solar energy, Georges Bataille (1988: 28) writes in the late 1940s, is both the "source of life's exuberant development" as well as the "origin and essence of our wealth." The sun is a giver and we are its receivers. Bataille (1988: 28) goes on: "Men were conscious of this long before astrophysics . . . they saw it ripen the harvests and they associated its splendor with the act of someone who gives." Solar's generous gift is one of ceaseless and boundless emissions that can never be returned. Over a half century later, Hermann Scheer (2002: 62) harnesses this imaginary to naturalize a new solar economy: "The sun will continue to give of its energies to people, plants and animals." This gift can power anything imaginable, Scheer writes, and will "satisfy even the most opulent energy demands." The generosity then cascades. Trudi Lynn Smith and James Rowe (2017: n.p.) speculate, "The sun's largesse can also inspire (re)surgent economic forms that are powered by generosity instead of avarice." These imaginaries are not only inspired by the possibility of energy abundance, but the potential "liberation of humans from earthbound resources" (Lorenz-Meyer 2017:

The South Atlantic Quarterly 120:1, January 2021
DOI 10.1215/00382876-8795668　© 2021 Duke University Press

431). This classical solar subject is a receiver of broadcasts, asked to tune in to aerial transmissions and to participate via consumption of invisible rays. Although distributed around the world, each located in a particular region and marked by difference, solar subjects become a mass through the reception of sunlight.

Just as all knowledge is generated from particular perspectives, all theory is generated out of a particular set of ecological contexts. Melody Jue (2020) proposes milieu-specific analysis as a means of taking into account how the environment, and one's position in relation to that environment, shapes the kinds of theories that result. A milieu-specific analysis of the solar economy reveals a particular kind of subject at the heart of these conceptions. It is a subject defined by visual capacities. It is a subject that watches plants grow. It is a subject that is called to look differently at the sun. It is a subject that is made responsible for being a different kind of receiver. And it is a subject inflected by the historical contexts, and media contexts, in which the solar economy gains traction. Scheer's argument for a decentralized solar future, originally published in 1999, is permeated by the ethos of the internet, but instead of free labor, the future is built on free energy. Resonating with the embodied human perception of the sun, this imagination of the classical solar subject and the generous sun is powerful. It hails its viewers, often quite consciously, to adopt a new subject position and, with it, an array of technologies and products that could precipitate a new energy future.

It would be possible to critique this solar subject, to document its emergence in the twentieth century and its crystallization as a networked entity at the dawn of the twenty-first, but this article is not invested in undoing what has been a successful catalyst for energy transition. Nor is it my intention to impede the energy transition itself, or the ways that, as Sheena Wilson describes, "Taking leave of oil as our main energy source could provide opportunities to develop more socially just ways of living that put the concerns of those most exploited—women, people of color, and the global 99 percent—at the core of energy transition politics" (2018: 378). Rather, I open by tracing these contours in solar theory—the orientational grounding of the classical solar subject—in order to open up some alternative ways of being and sensing solar.

The bulk of this article is an exploration of how solarity would look from other subject positions. What if we, specifically as the readers of this special issue and as those invested in knowing or researching solarity, are not figured as the sun's primary receivers? What if the visual, and sunlight as a condition of visuality, is not the primary mode by which the sun is experienced? What if the place we can most easily enact the transformations of solarity is not in harvesting the sun's radiant generosity, but in reconfiguring our relationship to the radiant regimes that remain invisibly embedded in the world around us? In other words, I ask questions that are intended to generate alternatives to the naturalized who, how, and where of solar theory, in order, as the Petrocultures Research Group (2019: 3) prompts, to offer "different ways of being in relation to one another and to the plurality of non-human others with whom our fates are entangled."

The central contribution of this article is to shift focus from directional solar rays, delivered by the sun, to what I call embedded solarities, the ways that solar energy, effects, and affects permeate the environment itself. It takes up the provocation that understanding solarity "demands that we attend to our relationships to materials, and to the infrastructures that mediate these relationships" (Petrocultures Research Group 2019: 5). Solarity, in this view, "is about turning to face these, as much or more than it is about turning to face the sun" (5). Looking at embedded solarities, it becomes clear that mediation is a vital part of solar materialities. Moreover, it is not simply that beings are sustained through the consumption of solar rays broadcast to earth, but that media work to reorganize radiant life, to redistribute life as it is entangled in a field of spectral exchange. In doing so, I take up Rahul Mukherjee's (2020) call to examine the "uncertain behaviors" of electromagnetic energy and radiant infrastructures.

This essay follows two sites of spectral, solar mediation in agricultural practice: the work of managing the sun in chicken reproduction and the negotiation of shade for agricultural laborers. While plant growth often provides the evidence of solar generosity for theorists, here the management of radiance reveals how sunlight can be weaponized through architectural and social formations as easily as it can be made beneficial. Moreover, growth is not simply a gift or effect of the sun, but is also mediated labor. As a result, the focus on embedded solarities affords a view of the distinct labor politics of solar energy—one that draws attention to its collective, exploitative and racialized dimensions rather than its individual potential.

If solar metaphors have at times endorsed solar's ability to ease the condition of Western urban and suburban subjects, this article offers another way of engaging with solar, one that foregrounds the embodied work that is required to benefit from sunlight. These agricultural practices reveal that it is not the sun that produces growth: growth—in a post-natural landscape—is always managed. Some forms must be suppressed and some species must be deprived, such that others can grow. This article is a call to track solarity beyond the sun, to turn to the ground and the body, to document manifestations of being solar across social practice and infrastructural assemblages, and human and non-human life. This will not only reveal the latent solar regimes that structure our varied inhabitations of the world, but can cultivate the forms of solar engagement that extend in and through the environment around us, even in the absence of access to solar technologies.

The Radiant Life of Chickens

In *Terror from the Air*, Peter Sloterdijk (2009: 91) describes the fundamental importance of atmosphere—and its mediation of solar rays—to life:

> If the earth, as parasite of the sun, came to be the birthplace of life—it draws just under a billionth of the sun's radiating energy—it is because water vapor and greenhouse gasses in the earth's atmosphere hinder the reflection of the sun's emitted short-wave energy in the form of long-wave infrared radiation, causing the earth's surface to warm to a life-compatible median temperature of plus 15.0 Celsius. . . . Life as we know it is contingent on the fact that, thanks to its atmospheric filter, the earth's surface lives thirty-three degrees beyond its means.

For Sloterdijk, life is not an outgrowth of the sun. It is an accident, "a side effect of having been climatically spoiled" (91). The atmosphere is a complex mediation of sunlight—filtering, trapping, and reflecting. It is because some light is stopped that growth itself could occur. So, too, are there a multitude of solar media on earth that manage the sun's spectral transmissions toward the production of life. Some of these amplify. Others do the work of filtration. In agriculture, a culture of managing such spectral exchanges, solar media are essential. While those who perceive it from a distance might observe that many farmers, even those who raise livestock, intimately and ultimately depend on the cultivation of plant matter through manipulated photosynthesis, even these practices depend on modulation rather than pure exposure.

Through agricultural practices of modulation, solar regimes—and the cultivation of radiant life—extend in and through the world. The multitude of objects (and food) that people are in contact with in everyday life thus reflects solar practices. In this section, I dive into one of the seemingly banal objects: a chicken's egg. It is perhaps the banality of the egg itself in Western digital cultural practice that suggests why it was chosen and clicked on enough to become the most-liked post on Instagram. Situated on a white background, divorced from context, this image of an egg exemplifies the

Figure 1. An egg, the most-liked post on Instagram, with over
fifty million likes as of February 2020.

relationship many Western urban subjects have had to their conditions of
bodily existence (Figure 1). It is an object of consumption through mouth
and/or eyes, is divorced from the vast webs of production that it enfolds, and
is globally circulated and endlessly discussed.

The production of an egg, just like the production of any food, embeds
a long history of social practices and infrastructural organization: from the
dispossession of land from its indigenous owners, to the construction of
massive infrastructural cold chains, to investments of the state in the organization of staple foods, to the regulation of economic transactions. Even on a
farm, a single egg embeds an extensive array of embodied movements and
technological practices. For example, for any small-scale egg farmer, daily

work likely involves retrieving, carrying, cleaning, cooling, organizing, counting, transporting, and cooking eggs. Hens lay, on average, between three and seven times a week, and each of these eggs in turn enfolds what Shane Brennan (2017) argues are "practices of sunlight": modes of mediating the sun through specific cultural—in this case agricultural—practices.

An egg enfolds a practice of sunlight because chickens are extraordinarily sensitive to light. Whereas people have three cones in our eyes, they have four. Many people can see about only 40 percent of what chickens can see: the latter can see in far more detail. Chickens' eyesight itself is formed through light exposure. As one farmer myth has it, before hatching, a chick turns in the shell. Its left eye is tucked into its body. Its right eye is next to the shell, and absorbs light through it. Exposed to light at this early stage, the story goes, the right eye becomes slightly nearsighted, allowing the chicken to hone in and search for food on the ground. The left eye, in the dark, becomes slightly far-sighted, enabling the chicken to detect predators from far away. Although there are no scientific studies on this, it is true that chickens can see the ways that small insects and seeds reflect ultraviolet light, and this makes it possible for them to identify and consume entities other than what are provided by a given farmer. Their ability to forage and how a given egg tastes corresponds to their sun exposure. Their reproductive cycle, and frequency of laying, is directly connected to sunlight. Many hens stop laying in the winter, when the sun exposure dips below fourteen hours a day. Massive factory farms use largely fossil fuel-based artificial light all year, and so do some small farms. But too much light, light overnight and without stop, produces agitation. Overexpose chickens to a sun substitute and they pluck their own feathers out and cannibalize each other. They overdose on light.

The chickens wake up the instant the light comes. And on small scale, free-range farms they roam until the light disappears. The architectures that keep them safe in the darkness, out of reach of nighttime predators, can intensify the sun's rays and their heat exposure during the day; as a result, farmers are up at dawn to let them out. The lives of these small chicken farmers are set to the rhythm of the sun. They rise with it every day and often wait on the animals as the sun sets. It is why, as farmers check social media at dusk, they see other farmers posting anecdotes from the day as they wait for their animals to head in. This kind of farming is just one of many cultures of solarity (and in many farming communities, social practices, meetings, and exchanges occur during sun-defined periods). This kind of farming is a way of living, working, and being that is shaped by the variations of sunlight, in which bodies move in response to sunlight. It involves many practices of sunlight, from the modulation of windows and shade, to the amplification of sun via plastic, that are engineered in response to the animals' and plants' light-sensing capacities. From this perspective, solarity is not about harnessing an energy source, it is about modulating a form of radiation, creating a series of prisms to refract and manage radiation into another being's growth and reproductive cycles. Through sun-mediation, farmers work on chicken reproduction; as a result, farmers can harvest eggs.

The consumption of eggs entangles people with radiant life, with solar media, and with practices of sunlight. It is difficult to see these embedded solarities in the egg itself. But just as bodies are transcorporeal (Alaimo 2010), extending beyond the limits of skin to the ecologies that sustain them, so too are the "objects" that people believe they hold in their hand or view on their screens. The chicken is a being that, in its sensitivity—not simply in its growth—is a form of radiant life that emerges with the sun and extends particular solar regimes. Industries and infrastructures enfold solar relations, and not just in their energy use, but also in their architectures, their means of organization, and their sales models. As Michelle Murphy points out, life itself is "a kind of varied enmeshment and enfleshment in infrastructures" (2017: 498). As a result, a legibility of infrastructure, supply chains, and industrial geographies is a prerequisite for seeing embedded solarities, the multitude of solar relations in which humans and non-humans are enmeshed.

Beginning this analysis of solarity with other beings, whose bodies and energetic production are cultivated in relation to sunlight such that people might consume them, allows other kinds of solar subjects to emerge: those who are already radiant, and whose bodies are managed by solar media in processes of spectral modulation. Solarity, in this analysis, is not limited to individual or even regional responsibility to tap into "free" rather than extractive energy, or to adopt sunlight systems rather than oil systems more broadly. Rather, an inquiry into the invisible and embedded regimes of solarity reveals that solar politics can materialize even in the everyday sites of food consumption, in animal welfare advocacy, and in many other social contexts that otherwise might be seen as outside the scope of a solar transition.

The Thermal Violence of the Sun

Much of solar theory stems from the human encounter with the sun as a light-making entity. The sun is described as casting rays of light. These are often described are directional transmissions, which parallel the illuminating beams of stage lights, cinema lights, and other artificial forms. This light not only communicates importance but also contains religious and spiritual dimensions. It has been critical as a figure of communality, "harmonious relationships and plenitude," Rhys Williams observes (2019: 11), in the speculative futures of solarpunk imaginaries. And light equates to visibility. In its dictionary definition, light is described first and foremost in relation to human vision: light is "something that makes vision possible" and "the sensation aroused by stimulation of the visual receptors" (*Merriam-Webster Dictionary* 2019). As a result of these correlations, much research has critically analyzed solar power in relation to light and dark. Scholars have charted specific cultures of light usage and described how the production of particular kinds of light, especially solar-powered lights, alters social stratifications and regimes of visibility. As Ankit Kumar's (2015) study of electricity in rural India reveals, light can signify hospitality. Similarly, Shane Brennan's (2017) study of the Detroit project Soulardarity shows that solar streetlights shape possibilities for "infrastructural visuality."

But thinking about sunlight, and what the sun offers, only in terms of visible light obscures a multitude of other radiant emissions. Many of the sun's rays remain beyond human perception. The entanglement of chickens with the sun doesn't just occur through the eyes. They sense the sun through the pineal gland in their brains. A blind chicken knows when it is daylight. The sun emits waves across the electromagnetic spectrum. And the sun is not the only radiator. The Earth is also radiant, bodies are all also radiant. They give off waves in the electromagnetic spectrum—free energy. These are other forms of light: invisible light, ultraviolet, and infrared. However, because the sun is typically thought of in terms of its visuality, these emissions often remain unaccounted for, even as they play a critical role in solarity's social formations. As one example of this, take the classic weather prediction interface, which offers indicators of heat, wind, and water via both precipitation and humidity, as well as sky cover—an index of sunlight. Collectively, and on the basis of proprietary algorithms, there is also an indicator for "Feels Like" temperature, which typically takes into account the role of wind in cooling and humidity in raising the felt temperature. The "Feels Like" temperature rarely includes solar radiation, and if it does, it does not include any difference between standing in the direct sun and sitting in the shade.

Such predictive or retrospective temperature reports are critical not only for farmers, but also for the many workers and people whose labor and safety depend on thermal regulations; notably, "true" temperature readings are always taken in the shade. The variability of solar radiation remains absent from such felt and institutional analytics and this produces a loophole for overheating and thermal violence to be enacted through direct sun exposure. In agriculture as a culture of solarity, the presence and absence of the sun is one of the most critical environmental indicators. The sun can actually change the felt temperature 10–15 degrees Fahrenheit and intensify heat. The sun—not simply temperature itself—can dry up the field and can burn out the weeds in between rows. The presence of the sun correlates to the amount of water that needs to be distributed, hoses extended, and row cover moved. It can be simply, and non-electrically, mediated through plastic, through paper, and through wood, among many other things, and it can be in used in turn to alter movement, water flow, and biological growth.

Beyond farming, sunlight is a resource always refracted by architectures and social practice. It is not equally accessible. Many solar advocates acknowledge this even as they talk about the abundance of the sun. They argue that a transition to a new energy economy will, as a result, require a mix of generation technologies and a regional, or even local, approach to energy production. Scheer (2002: 67) writes that it is the "great structural variety"—the fact that energy will be produced differently in different places—"which makes it difficult for energy ministers, who for decades have been used to the structures of fossil fuel supply, to get a feel for the potential of renewable energy." Solar here is a means of decentralizing energy and shortening the supply chain, or more radically, producing "electricity with no supply chain" (Scheer 2002: 77). It is on the basis of potential decentralization that solar requires a "radical rethink of the supply and distribution network" (Scheer 2002: 75). Even if the sun is universal, the capacity to access and harness it is highly local and spatially mediated.

At one extreme, take the fact of differential access to sunlight. After the South Carolina Department of Corrections installed metal plates to block sunlight from cells, prisoners began to spread the hashtag: #SunlightIsaHumanRight. On the other coast, in Los Angeles, shade is the scarce resource. Here, a form of thermal violence is enacted through architectures, city planning, and regulation. Practices of sunlight make it difficult to find shade, especially for marginalized black and brown communities in the city. Sam Bloch's (2019) study of the city describes how "sunlight was weaponized" to "clear out" such populations. Outside the city, farmworkers labor in what Edward R. Murrow once called "Sweatshops of the Soil" (Friendly 1960). California farmworkers are covered with hats, bandanas, long sleeves, and pants to protect themselves from the sun. California has passed heat illness laws, requiring farms to provide shade for their workers, but this hasn't stopped agricultural enterprises from killing farmworkers in the fields. In the United States, heat stroke from laboring in the hot sun is a leading cause of death for these workers (Center for Disease Control and Prevention 2008).

Just as an egg enfolds a set of practices of sunlight, or energy-dependent artificial lighting, so too do the vast majority of fruits, many of which still require people to pick them by hand in the sunshine. Pointing out the extractive zones upon which the production of solar technology depends, as well as the large-scale solar enterprises that exist today, the Petrocultures Research Group (2019: 3) says that "solarity might be continuous with the capitalist, masculinist, racist, colonialist and imperialist extractive enterprises that have defined the fossil-fuel era globally." Many practices of solarity already are continuous with these enterprises: the growth of biomatter for the consumption of remote publics requires not only a negotiation of the sun, of technologies of (often plastic) sun amplification and resistance, but also the mobilization of bodies in and through solar forms of production. Through some peoples' overexposure to the sun, others benefit.

In short, it is not simply that sunlight is mediated and differentially accessed, but that it conveys much more than light. Its heat is less often accounted for and aligned with descriptions of the sun's generous gift. Were scholars to refocus on the sun as a source of infrared rays, it would be clear—as Sloterdijk describes—that the capacity to filter is as critical to stimulating growth and maintaining forms of radiant life as capture is. In turn, the capacity to amplify sunlight can easily be turned into thermal violence. Practices such as sweatboxing—a practice of intentionally overheating the body—have long mobilized the sun as a means of enacting harm (Starosielski 2019). Turning to embedded solarities, then, involves looking at the layered modulations of solar radiation as not only a productive force, but also as a destructive one that often defers accountability to the environment itself.

Beyond the Sun

Instead of thinking of sunlight as a natural, democratically, and universally available resource—one that is evident through vision—all of the examples I've provided here prompt a different line of thinking about sunlight: as both a socially mediated resource and a potential weapon. Attuning to embedded solarities is a means of investigating the differential access to and effects of solar exposures. Attuning to radiant life directs attention to the ways that bodies are comprised through spectral manipulations.

These analytics prompt a logistical and infrastructural understanding of solarity in which radiance is not limited to a relationship to the sun, but can be tracked to and through the spectral formations in which bodies are enmeshed. Identifying solarity's possibilities and potential politics thus requires first looking out to the world to parse solar's embedded forms in infrastructures and invisible spectral connections. It requires an active understanding of the effects of solar regimes and the intricacies of their production. Scholars often marvel at the sublimity of phenomena beyond the scale of the human—whether infrastructures, supply chains, or networks—and their vast and expansive reach. While it is difficult to conceive of an individual grasping the entirety of expansive networks of logistical coordination, it is possible for people collectively to engage sites of embedded solarities. It is possible to document where sunlight becomes a form of communication, how different kinds of growth enfold the sun, and how bodies are overexposed in everyday labor. To restructure what research looks like in relation to solarity is thus to engage beyond-the-human-scale problems of infrastructure differently.

How might people take such observations to move toward an energy transition animated by solarity? How can solarity be a new form of life, of production of life, of living with the sun? I end here by suggesting that critical attention to embedded solarities can form the basis for a set of more equitable ways of working with and in relation to the sun, and not just as an energy source, but also as a part of everyday environments and inhabitations. Solarities can be more justly engaged in activism around food systems, for example, in which techniques of solar modulation are assessed and accounted for. Solarities can be more justly engaged in work around prison abolition, and through the identification of carceral operations that limit one's capacity to modulate sun exposure—whether in forced exposure or in deprivation. Solarities can be engaged in everyday life by experimentation, multiplication, and distribution of technologies of sun-mediation—from architectures and clothing to everyday social practices. Each of these examples is a site where new possibilities for an energy transition and life with the sun are already embedded in the work for justice on the ground.

References

Alaimo, Stacy. 2010. *Bodily Nature: Science, Environment, and the Material Self.* Bloomington: Indiana University Press.

Bataille, Georges. 1988. *The Accursed Share: An Essay on General Economy.* NY: Zone Books.

Bloch, Sam. 2019. "Shade." *Places* (April), placesjournal.org/article/shade-an-urban-design-mandate.

Brennan, Shane. 2017. "Practices of Sunlight: Visual and Cultural Politics of Solar Energy in the United States." PhD diss., New York University.

Center for Disease Control and Prevention. 2008. "Heat-Related Deaths among Crop Workers 1992–2006." Atlanta, GA. Center for Disease Control and Prevention.

Friendly, Fred W. 1960. "Harvest of Shame." *CBS Reports*, CBS, November 25.

Jue, Melody. 2020. *Wild Blue Media: Thinking Through Seawater*. Durham, NC: Duke University Press.

Kumar, Ankit. 2015. "Cultures of Lights." *Geoforum* 65, October: 59–68.

Lorenz-Meyer, Dagmar. 2017. "Becoming Responsible with Solar Power? Extending Feminist Imaginings of Community, Participation, and Care." *Australian Feminist Studies* 32, no. 94: 427–44.

Merriam-Webster Dictionary. 2019. "Light." merriam-webster.com/dictionary/light.

Mukherjee, Rahul. 2020. *Radiant Infrastructures: Media, Environment, and Cultures of Uncertainty*. Durham, NC: Duke University Press.

Murphy, Michelle. 2017. "Alterlife and Decolonial Chemical Relations." *Cultural Anthropology* 32, no. 4: 494–503.

Petrocultures Research Group. 2019. *Solarity: Energy and Society after Oil*. After Oil School II. Montréal, Québec, May 23–25. afteroil.ca/solarity-energy-and-society-after-oil/.

Scheer, Herman. 2002. *The Solar Economy: Renewable Energy for a Sustainable Global Future*. London: Earthscan.

Sloterdijk, Peter. 2009. *Terror from the Air*, trans. Amy Patton and Steve Corcoran. Los Angeles: Semiotext(e).

Smith, Trudi Lynn, and James K. Rowe. 2017. "Pipelines as Sun Tunnels: Visualizing Alternatives to Carboniferous Capitalism." *CTheory*. ctheory.net/ctheory_wp/pipelines-as-sun-tunnels-visualizing-alternatives-to-carboniferous-capitalism/.

Starosielski, Nicole. 2019. "Thermal Violence: Sweatboxes and the Politics of Exposure." *Culture Machine* 17. culturemachine.net/vol-17-thermal-objects/thermal-violence/.

Williams, Rhys. "'This Shining Confluence of Magic and Technology': Solarpunk, Energy Imaginaries, and the Infrastructures of Solarity." *Open Library of Humanities* 5, no. 1: doi.org/10.16995/olh.329.

Wilson, Sheena. 2018. "Energy Imaginaries: Feminist and Decolonial Futures." In *Materialism and the Critique of Energy*, edited by Brent Ryan Bellamy and Jeff Diamanti, 377–412. Chicago: MCM Publishing.

Dominic Boyer

Revolution and *Revellion*:
Toward a Solarity Worth Living

Adaptation to three, four, not to speak of eight degrees
is bound to be a futile endeavour. No matter how
advanced the sprinklers Syrian farmers install,
irrigation requires water. No walls can save the Nile
Delta from the underground infiltration of the sea.
No one can perform any kind of physical labour when
temperatures settle above a certain level, and so on.
But the proven fossil fuel reserves can be kept in the
ground. Emissions can be slashed to zero. "Everybody
says this. Everybody admits this. Everybody has
decided it is so. Yet nothing is being done," and this is
the rationale for the most exigent type of revolution,
the one that, in full consciousness of the roots of the
problem, wages a full-scale onslaught on fossil capital,
just as the Bolsheviks set themselves the task of putting "an immediate end to the war," insisting that
"it is clear to everybody that in order to end this war,
which is closely bound up with the present capitalist
system, capital itself must be fought."
—Andreas Malm, "Revolution in a Warming World"
(2017)

Play captures a lot of what goes on in the world.
There is a kind of raw opportunism in biology and
chemistry, where things work stochastically to form
emergent systematicities. It's not a matter of direct
functionality. We need to develop practices for thinking
about those forms of activity that are not caught by
functionality, those which propose the possible-but-
not-yet, or that which is not-yet but still open.

The South Atlantic Quarterly 120:1, January 2021
DOI 10.1215/00382876-8795682 © 2021 Duke University Press

It seems to me that our politics these days require us to give each other the heart to do just that. To figure out how, with each other, we can open up possibilities for what can still be. And we can't do that in a negative mood. We can't do that if we do nothing but critique. We need critique; we absolutely need it. But it's not going to open up the sense of what might yet be. It's not going to open up the sense of that which is not yet possible but profoundly needed.
—Donna Haraway, "A Giant Bumptious Litter" (2019)

This collection acknowledges that the world is experiencing a time of profound, even civilizational, collapse, and rebirth. The modern civilization (modernity) that began the twenty-first century was built upon two fossil fuel regimes—the coal regime that permitted an abundance of engines, machines and electricity and the oil regime that allowed engines and machines to colonize landscapes, skies, and seas to an unprecedented degree. It has been convincingly argued that fossil fuels inherited and accelerated the extractive legacies of colonial frontiers (Moore 2015; McNeish and Logan 2012; Mouhot 2011), in which Europeans sought greater magnitudes of productivity and efficiency than could be coaxed from the killing fields of plantation and mine labor. It has also been convincingly argued that fossil fuel logics are fundamental to modern capitalism and are reflected for example in economic norms of growth and consumption (Kallis 2018; LeMenager 2013). Coal and oil have quite literally been the engine of modern capitalist relations since the eighteenth century (Huber 2013; Malm 2013; Szeman 2007). There are lively debates as to just how much longer those relations can persist without provoking widespread ecological collapse—ranging from not a minute longer to roundly arbitrary dates like 2030, 2050, and 2100—suffice it to say that if the twenty-first century ends the way it began then it may very well be humanity's last.

In fits and starts, post-fossil ways of life are beginning to take shape. But when we try to imagine what kind of modernity might come next, we immediately feel the weight of historical inheritances limiting a sense of possibility. Looking across popular culture, it has obviously proved easier to imagine the post-fossil world in the negative register of apocalypse or ruins than to positively imagine the widespread organization of stable non-capitalist, non-extractive, non-fuel-driven relations. In this collection of essays, the contributors seek to contribute to new imaginative horizons by offering the gloss of "solarity" to capture the world that is now becoming. But what is solarity exactly? What are its qualities and conditions of possibility? In the first place, it is helpful to consider solarity less as a precise political program than as the process of negating across large and small scales the reproductive apparatus of fossil-fueled modernity (petromodernity). In his early writings, Marx described "communism" in similar terms, e.g. not as a clearly defined societal alternative to bourgeois relations but rather as the dialectical project of negating bourgeois alienation and transcending it. Good Hegelian that he was, Marx was doubtful that the positive forms of post-bourgeois sociality would be knowable to people raised within the logic of the ancien regime. Marx was, if you will, form-agnostic. In my opinion, solarity ought to be agnostic to its future forms as well. It's difficult to think one's way into a fully realized post-fossil world right now, but that imaginative leap will become less daunting as fossil logics and institutions are negated more widely.

However, being agnostic as to the exact forms solarity might adopt does not mean not caring about the values that would inform a solar modernity. If, for example, rampant consumerism, self-indulgence, and environmental neglect have been hallmarks of petromodernity, then we should value more highly human-nonhuman relationality, care, and humility in our solar futures. And, above all, we should hope that the social possibilities that negating fossil-fuel inheritances will open up would be excessively multiple. History has shown that utopian platforms that tolerate only one way of being in the world tend to be those associated with the greatest violence and terror.

Join the Revolution?

Solarity certainly sounds like a call to revolutionary action. The work of this essay is to discuss whether revolution ought in fact to be the language of imagining and enacting solar futures. Revolution has an impressive historical resume to be sure. But it is also increasingly a discursive readymade of the petrocultural status quo. A quick foray into Google will net you cyberbuskering about the production revolution, the resource revolution, the digital revolution, the AI revolution, the payment revolution, the transportation revolution, the fourth industrial revolution and many more revolutions besides. Dig a little deeper and you'll find elaborate "Join the Revolution!" corporate marketing campaigns. A3 Performance wants you to buy revolutionary swim gear, All Elite Wrestling wants you to buy their revolutionary helmets and ringtones, SodaStream—until recently best known for their exploitation of Palestinian labor—is now offering a "disruptive recruitment video" that promises a revolution (in what it isn't clear but no mention is made of helping Palestinians), even Weetabix—Et tu, Weetabix?

"Join the Revolution!" has obviously become the opiate of masses of uninspired twenty-first-century ad writers. This is not to belittle other more earnest contemporary calls to revolutionary movements and actions, whether in the name of fomenting a climate revolution, a renewable energy revolution or a decolonial revolution. My point is just that the majority of revolution talk circulating in public culture these days is quite unrevolutionary from the point of view of actually demanding substantive social change.

It is tempting, of course, to blame the dilution of revolutionary discourse simply on the plague of capital, which makes of every idea a commodity and for every commodity a market. But the trouble with revolution talk goes deeper. Hannah Arendt observed for example that, at least until the French Revolution, the political meaning of *revolution* continued to be overlaid conceptually with its astronomical reference of "recurring, cyclical movement." Revolution could thus mean the restoration of monarchical authority as in the case of the Glorious Revolution, a far cry from the modern ideal in which revolutionary actors "are agents in a process which spells the definite end of an old order and brings about the birth of a new world" (Arendt 1963: 32).

The newness of those modern worlds has to be questioned as well. Even though the famous revolutions of the late eighteenth and early nineteenth centuries promised novelty by centering values of freedom and liberty (and lest we not forget private property) against regimes of aristocratic inheritance and privilege, it is no secret that liberal political philosophy contained its own internal hierarchies of what classes of persons were judged deserving of freedom. The revolution of that era that imagined freedom in the most expansive way—Haiti—has been brutally punished for its audacity ever since by being subjected to what is perhaps the longest campaign of odious debt bondage in human history (Buck-Morss 2000; Trouillot 1995). Liberal revolutions left vast political hierarchies intact, championing the virtue of freedom on the one hand while allowing genocide, slavery, patriarchy, class inequality to freely persist and reproduce on the other.

Ariella Azoulay captures this contradiction precisely when she writes that revolutions, both in concept and in practice, are not civil. That is to say, revolutions tend to take the form of minoritarian regime change enacted through violence and thoroughly opposed to "the civil power of the many and threatening from the start to prevent its constructing alternative formations of congregation, partnership, and sharing" (Azoulay 2012). For Azoulay the epitomizing revolutions of the eighteenth century, the American Revolution and the French Revolution, "enabled an anti-monarchy, white-male minority to obtain a better-ordered domination of the masses and *distill out of the body politic they had created the citizens whose right to self-determination and sphere of agreement became the very heart of revolutionary legacy and regime.*" One might say that the epitomizing revolutions of the early twentieth century—the Russian and Mexican Revolutions—eventually resembled changings of the masculinist elite guard too. This is not to underestimate the potential of socialist revolution to unsettle inherited political hierarchies and to encourage certain kinds of civil power. The revolutionary Mexican Constitution of 1917 created at least the theoretical possibility of indigenous sovereignty even if that promise has scarcely been acknowledged let alone enacted by the political regimes that followed it. And, as someone who began his career researching the legacies of the German Democratic Republic, I know from hearing countless personal testimonies that Stalinized socialist societies were capable of creating opportunities for gender equality and class mobility far more effectively than their liberal-democratic counterparts.

Another problem trails not only the early twentieth-century revolutions but also most decolonial revolutions of the mid-twentieth century as well. For whatever else liberal, socialist, and nationalist movements disagreed about, they aligned remarkably well around the idea that their best path forward was through high-carbon modern industrial and urban development (Boyer 2016). One sees this spirit in Lenin's redefinition of communism as "the Soviets plus electrification," in the petronationalisms of Latin America, in the Islamic petromonarchies of the Middle East, and in the many newly independent African nations convinced to build programs of social development around oil exports in the 1950s and 1960s. Even those outlier countries who committed to relatively low-carbon models of electricity provision (Brazilian hydropower, French nuclear power) did not evade committing to other kinds of high-carbon path dependency ranging from automobility and air travel through to reliance upon emissions-intensive building materials like concrete. One could possibly credit the Maoist agrarian revolutionary movements of the twentieth century with divergence from this mainstream pathway. But to credit the genocidal practices of, say, the Khmer Rouge with decarbonization is obviously perverse and only underscores Azoulay's argument that revolutionary violence is anti-civil in the extreme.

So where does this leave us as we contemplate the need for radical climate action today? Can the concept of revolution be restored in a way that fits the urgency of the moment without repeating the violent tragedies of the past? Azoulay offers the horizon of "civil revolution" as an alternative to the historical record:

> Civil revolution means correction, reparation, repartition, imagination, common experience, possible dreams. This is a language spoken by individuals in different places in the world. When they have had enough of the sovereignty of the nation-state and the capital to which they are subjugated, enough of the evil it produces and its oppression of them and others in the shadow it casts over the horizon of imagination, their gaze, speech, and action, they begin to speak it in public. They seek interlocutors, rubbing against others who speak as they do and resolve to speak with each other in civil language, no matter what. Urgency drives them to imagine and to act, doing so not behind closed doors but rather in the presence of others—foreigners and strangers—like them. (2012)

This horizon of civil revolution is a wise intervention, just as there is much to recommend the rationale for revolution that Andreas Malm offers in the epigraph to this essay. A concept is only as good as its application, of course. If what is sought were a truly novel world of better relations, more care, more equity, and more peace then I would happily join a revolution to attain that world. Under the right circumstances and with the right voices, there is every reason to maintain a language of revolution if it is specific and motivated and not a leaky dangerous vessel of capitalist marketing or violent fantasy.

But it is also worth saying that the obvious contradictions of revolution talk are perhaps not a tragedy. We could greet the often empty and even grotesque invocations of revolution today as reminders that there is still important work to be done to find the apt concepts and narratives for describing radical change at this unprecedented moment in the history of the planet. Every era deserves its own utopias, perhaps especially solarity. But we have already seen how old high carbon utopias have sought to mutate and capture the future, putting forth paradoxical beacons like *sustainable growth* and *green capitalism* to inhibit the emergence of truly post-fossil imaginaries. The utopia of radical change I would like to put forward here is one that subverts the industrial modern distinction between work and play.

Work and Play

Many of us who have written on climate action have described it as hard work. The narrative language is one of a long arduous struggle to end the ecocidal trajectory of fossil civilization. To take just one example, in his introduction to Extinction Rebellion's *This is Not a Drill* handbook, Sam Knights

(2019: 9) describes XR's path as "long and arduous" and "hard and punishing," and predicts that even the act of reading the handbook is sure to inspire "feeling sad, or empty, or guilty, or angry, or frightened, or numb" (13). To be fair, Knights also writes about the positive togetherness of the movement, about learning to dream together, and even rewilding the imagination. But as is so common in reflections on radical change today, the overwhelming emphasis is on civilizational change as a matter of exhausting labor.

Of course, this is not without reason. The phenomenological experience of climate action is one of constant struggle because of the sheer density of petro-infrastructures (from forms of knowledge to commodities to mobilities) with which one interacts on a daily basis. Negating petroculture means negating much of everyday life in the Global North (and much of the Global South): one would ideally avoid automobiles and airplanes, avoid the products of industrial agriculture, avoid energy-intensive and single-use commodities of all kinds, avoid coal-powered electricity, avoid the use of petroplastic materials, and, of course, avoid naturalizing ideas of consequence-free growth, travel and energy use. All that avoidance requires concentration, discipline and is indeed difficult (as is suffering the moralizations that follow from an individual's failure to be consistently and properly ascetic).

An irony that is much more than an irony is that were all of us in the global North suddenly to decide to work less hard, our emissions intensity would decline very likely more rapidly than were we to work harder at being greener. Is this not the lesson of the first few months of the coronavirus pandemic, which has created clean skies across the world? This is controversial. But in fact, doing less in general (including consuming less, travelling less, mediating less, working less) would be easier to achieve than maintaining the same levels of energy expenditure while decarbonizing energy sources. I have in mind also the parable of Marshall Sahlins's "original affluent society" (1972) that if productive activity is considered more as an occasionally necessary evil rather than a permanent obsession then leisures and pleasures become less an anxious vanishing point—commodities ever sought and rarely enjoyed—and more the norm of lived experience.

Modern conceptualizations of work and energy also co-relate. In Cara Daggett's (2019: 86) brilliant reconstruction of the origins of the modern sense of energy as physical work—a sense much less capacious than Aristotle's original *energeia*—she describes how the logic of energy became infused with "Protestant inclinations" that positioned work as a form of virtue: "Energy helped to adapt the work ethic to a technoscientific era. The energetic model of work buttressed work's cultural value in the West—already high—with a new scientific justification and language." Joseph Campana (2017: 65) writes of how the centralization of oil capitalism in modern economies infiltrated cultural rhythms, creating an "interlacing of energic and affective cycles constituted by the oscillation between booms and busts" and manifesting in wild swings of exuberance and catastrophe. For Campana (65), petroculture disposes us to manic behavior even in our efforts to escape petroculture; he instead urges us to resist our most zealous impulses and instead to explore "powering down," not so much in the sense of turning off lights and turning to bikes but rather by retraining "the susceptible and interlocking circuits of feeling and flesh" to do less.

But I think there is also good reason to imagine that the options for radical change are not limited to powering up and powering down. A third term that disrupts both the regime of modern work and its companion promise (more fantasy than experience) of leisure is the space of play. Donna Haraway, in the second epigraph to this essay, positions play as a necessary supplement to critical environmental practice. She uses the concept of play in a very specific way. For Haraway, play is certainly not the non-productive temporal waste that the western industrial work-regime constantly trivializes and militates against. Likewise, it is not the kind of market-mediated fun that easily becomes an alibi for industrial work-without-pay. Haraway's conceptualization of play does have certain resonances with Huizinga's proposition that play is a constant and pivotal feature of cultural order making. Play is generative for Huizinga (1949: 197), a space of freedom framed against extant cultural norms and yet provoking new cultural possibilities; he describes the "play-spirit" in opposition to "systematization and regimentation" and also describes the "precarious balance between seriousness and pretence [as] an unmistakable and integral part of culture as such" (191). Haraway's play is likewise a generative principle but she strengthens the ontological aspect of the argument. Haraway's play is an aspect of "worlding" (Stewart 2010) that she contrasts with "the kind of possessive individualism that sees the world as units plus relations. . . . People like me say, "No thank you: it's relationality all the way down." You don't have units plus relations. You just have relations. You have worlding" (Haraway 2019).

Haraway's play could thus be described as experimental relation making. Playing with the world to see what ways of thinking and being might be able to endure in a time of worldly crisis. Haraway focuses specifically the "not-yet" and "might-still-be" of emergent but not-fully-functionalized forms of political activity. She counsels, "It seems to me that our politics these days require us to give each other the heart to . . . figure out how, with each other, we can open up possibilities for what can still be. And we can't do that in a negative mood. We can't do that if we do nothing but critique. We need critique; we absolutely need it. But it's not going to open up the sense of what might yet be." (2019; also Haraway 2008, 2016) What interests me about this vision of play is that it seems very closely aligned with the mission of solarity. The question is what Haraway's sense of play might mean in the context of what must still be rightly and fully appreciated as the hard work of radical change.

Revellion

In the spirit of commitment to emergent political possibilities, those of us engaged in projects of solarity might lay greater emphasis on the often-unspectacular process of *de*systematizing the reproductive apparatuses of petromodernity. These are everyday practices of sabotage that seek to make rubble of the material and epistemic infrastructures of high carbon life. Giving up a personal automobile for a bicycle is an act of sabotage; collaborating with neighbors to create a local solar microgrid is another; the experience of low-carbon pleasure is still another. The tools are always ready at hand, because the tools are very often simply the reassertion of limbs and minds over fuel-powered machines. At this moment, such practices are the essence of solarity. Petroculture is a juggernaut that will have to be ruined in order that its elements can be made available to redistribute into new projects. Although desystematization may often be experienced as hard work, the scale of its practice and the relations of care that enable it mean that solarity is rarely abject and indeed often carries with it a sense of mutual joy. Some years ago, David Graeber described the neo-anarchistic activity of prefigurative politics in similar terms, a humble and collective finding forward:

> This is very much a work in progress, and creating a culture of democracy among people who have little experience of such things is necessarily a painful and uneven business, full of all sorts of stumblings and false starts, but—as almost any police chief who has faced us on the streets can attest—direct democracy of this sort can be astoundingly effective. And it is difficult to find anyone who has fully participated in such an action whose sense of human possibilities has not been profoundly transformed as a result. It's one thing to say, 'Another world is possible'. It's another to experience it, however momentarily. (2002: 72)

Desystematization is a cumbersome word and also very much a critical negation. I have been thinking about how the idea could be rearticulated as a positive mission. In doing so, I stumbled across a marvelous semantic slippage in medieval French between the verbs *reveler* (to be disorderly, to make merry) and *rebeller* (to revolt), which has allowed for an occulted kinship between the modern concepts of rebel and revel. So, is *revellion* a term that might speak to our times? Revellion, meaning a mode of insurgency that is playful—in both the experimental and ludic senses of play—a riot that is riotous. Revellion does not abide the petrocultural oscillation between seriousness and extravagance; it seeks somehow to be both at once.

When one speaks of introducing play into political projects, one frequently attracts criticism from more earnest revolutionaries that one is giving oneself over to the capitalist economy of pleasure and its desire to make a game of any serious political idea. As discussed above, such gamification does abound, and all the better to consign potentially threatening ideas to the domain of leisure time. But I think revellion means something quite different. I think it is about a kind of re-worlding of limbs and minds in order to encourage an excess of possibility. This was precisely the revellious spirit of Iceland's Best Party that I have documented at length elsewhere (Boyer 2013); Jón Gnarr and his collaborators demonstrated that a playful, experimental occupation of the domain of so-called serious politics was a highly effective strategy for opening new political horizons and for encouraging wider participation in the process of system change beyond those belonging to the political class. Importantly, it was also not a lasting occupation. Although the Best Party was leading in the election polls in 2014, it disbanded itself rather than seeking a second term in office. Gnarr (2014) commented, "the Best Party was a surprise party and you can't walk around a party that is already happening shouting 'surprise!'" He went on to say that he and his collaborators hoped that their dissolution would create the space for others, especially young feminists, to take their place in remaking politics. At the same time, the Best Party escaped being functionalized in the system of so-called normal politics. As Ivan Illich (1973: 40) once credited poets and clowns, "Their intimate wonder dissolves certainties, banishes fears, and undoes paralysis."

Nurturing such a sense of continuous possibility is vital for severing ingrained, fossil fueled habits of thought and action. Seriousness is often an enemy of solarity. Serious people frequently advise that revolutions (*real* revolutions) are impossible, childish fantasies. But a revellious spirit can outwit them because it refuses to play *their* game. What Haraway doesn't say exactly (but I suspect she is thinking it) is that an unplayful revolutionary could undermine solarity's better instincts. Whatever solarity we can imagine from our present political situation is a solarity that is bound to still be confined within certain functionalities of petroculture. It is not our fault, a century of petroculture has left its mark. With humility (also a core affect of XR by the way) we need to accept that we don't have everything figured out and that there are emergent political relations still to be discovered and embodied. We must take heart that the post-fossil world will come into sharper focus over time.

This is not to discount the importance of trying to scale better relations into new habits and institutions immediately. But rather to reiterate that the positive forms of solarity are not all available to us today. It will take both work and play to find them. Meanwhile, to imagine civilizational change solely in the key of work is to smuggle back in the Protestant energy-labor nexus that has contributed so much to the Anthropocene/Capitalocene trajectory of "improvement" at all costs. In our field research on energy transition in Southern Mexico, it was crippling to hear Indigenous *Binniza* and *Ikojts* peoples describing wind parks in the same breath as mines. Here was a green capitalist initiative that felt exactly like the settler-colonial predations and dispossessions of the past. One Indigenous activist told us, "Maybe we are seeing a transition in the forms of energy, but there is a clear continuity in the form of resource exploitation. . . . And so, I have to ask, What 'transition'? I don't feel that there is a transition. What change is there? There is no change here. Only talk. And I think that the discourse [about climate change] is being exploited as well. There is worldwide concern about climate change, and the companies are monetizing this as well" (Howe 2019: 112).

The contradictions within green capitalism are increasingly obvious. There is no green growth. That is just a shiny poison pill. The contradictions are tragic but they also remind us that we live in epic times. There has never been the need to remake global civilization before and we have to treat this as the unique opportunity it is to break the inheritances of white masculine supremacy that built the global capitalist order. What I am suggesting is an ethos of unmaking the Anthropocene trajectory that is less heroic and more attuned to the "hyposubjective" character of our present situation (Boyer and Morton 2020). We must always ask ourselves: Are we creating a solarity that is worth living? In bits and scraps, solarity is (be)coming and even if we don't quite know yet what forms it will take, we know that the road will involve a lot of playful work and workful play.

References

Arendt, Hannah. 1963. *On Revolution*. New York: Penguin.

Azoulay, Ariella. 2012. "Revolution." *Political Concepts: A Critical Lexicon* 2: politicalcon-cepts.org/revolution-ariella-azoulay/.

Boyer, Dominic. 2013. "Simply the Best: Parody and Political Sincerity in Iceland." *American Ethnologist* 40, no. 2: 276–87.

Boyer, Dominic. 2016. "Revolutionary Infrastructure." In *Infrastructures and Social Complexity: A Companion*, edited by Penelope Harvey, Casper Bruun Jensen, and Atsuro Morita, 174–86. London: Routledge.

Boyer, Dominic and Timothy Morton. 2020. "Hyposubjects." In *Anthropocene Unseen: A Lexicon*, edited by Cymene Howe and Anand Pandian, 233–35. New York: Punctum Books.

Buck-Morss, Susan. 2000. "Hegel and Haiti." *Critical Inquiry* 26, no. 4: 821–65.

Campana, Joseph. 2017. "Power Down." In *Veer Ecology: A Companion for Environmental Thinking*, edited by Jeffrey Jerome Cohen and Lowell Duckert, 60–75. Minneapolis: University of Minnesota Press.

Daggett, Cara. 2019. *The Birth of Energy: Fossil Fuels, Thermodynamics, and the Politics of Work*. Durham, NC: Duke University Press.

Gnarr, Jon. 2014. "What Happened?" *Reykjavík Grapevine*, May May. grapevine.is/mag/feature/2014/05/26/what-happened/.

Graeber, David. 2002. "The New Anarchists." *New Left Review* 13: 61–73.

Haraway, Donna. 2008. *When Species Meet*. Minneapolis: University of Minnesota Press.

Haraway, Donna. 2016. *Staying with the Trouble: Making Kin in the Chthulucene*. Durham, NC: Duke University Press.

Haraway, Donna. 2019. "A Giant Bumptious Litter: Donna Haraway on Truth, Technology, and Resisting Extinction" *Logic*, no. 9. logicmag.io/nature/a-giant-bumptious-litter/.

Howe, Cymene. 2019. *Ecologics: Wind and Power in the Anthropocene*. Durham, NC: Duke University Press.

Huber, Matthew T. 2013. *Lifeblood: Oil, Freedom, and the Forces of Capital*. Minneapolis: University of Minnesota Press.

Huizinga, Johan. 1949. *Homo Ludens*. London: Routledge.

Illich, Ivan. 1973. *Tools for Conviviality*. New York: Harper and Row.

Kallis, Giorgios. 2018. *degrowth*. New York: Columbia University Press.

Knights, Sam. 2019. "Introduction: The Story So Far." In *This Is Not a Drill: An Extinction Rebellion Handbook*, edited by Clare Farrell, Alison Green, Sam Knights, and William Skeaping, 9–13. London: Penguin.

LeMenager, Stephanie. 2013. *Living Oil: Petroleum Culture in the American Century*. Oxford: Oxford University Press.

Malm, Andreas. 2013. "The Origins of Fossil Capital: From Water to Steam in the British Cotton Industry." *Historical Materialism* 21, no. 1: 15–68.

Malm, Andreas. 2017. "Revolution in a Warming World." *Socialist Register 2017: Rethinking Revolution*. London: Merlin.

Marx, Karl. 1964. *Economic and Philosophic Manuscripts of 1844*. New York: International Publishers.

McNeish, John-Andrew, and Owen Logan, eds. 2012. *Flammable Societies: Studies on the Socio-Economics of Oil and Gas*. London: Pluto.

Moore, Jason W. 2015. *Capitalism in the Web of Life: Ecology and the Accumulation of Capital.* London: Verso.

Mouhot, Jean-François. 2011. *Des Esclaves énergétiques: Réflexions sur le changement climatique.* Paris: Champ Vallon.

Sahlins, Marshall. 1972. *Stone Age Economics.* New York: Routledge.

Stewart, Kathleen. 2010. "Worlding Refrains." In *The Affect Theory Reader,* edited by Melissa Gregg and Gregory J. Seigworth, 339–53. Durham, NC: Duke University Press.

Szeman, Imre. 2007. "System Failure: Oil, Futurity, and the Anticipation of Disaster." *South Atlantic Quarterly* 106, no. 4: 805–23.

Trouillot, Michel-Rolph. 1995. *Silencing the Past: Power and the Production of History.* Boston: Beacon.

Eva-Lynn Jagoe

Solar Goop:
Energy Futures and Feminist Self-Care

The thick clusters of roses are pale pink on the
outer edges and darker as they get closer to the
center, which is a bright deep red. Maybe ten
feet high, the almond-shaped arch beckons deli-
ciously, obscenely. It could be a surreal set in a
Pedro Almodóvar film, or some tamed version of
Allyson Mitchell's 2004 installation, "Hungry
Purse: The Vagina Dentata in Late Capitalism."[1]
Then Gwyneth Paltrow steps in front of it. You
wish she would get out of the way so you could
continue to gaze at the beckoning carmine center,
though her ivory-colored midriff-exposing outfit
sets her off to good advantage against the floral
backdrop. Her grin is exuberant.

Our task in this volume is to speculate on
what solarity looks like "in terms of social systems, asking questions not about technology but
about relationality and modes of being" (Szeman
and Barney, this issue). I take this to be an invi-
tation to imagine the different intimacies that
solar could afford, intimacies that have the potential to be radically different from the relational
structures afforded by petro-modernity. If we let
ourselves fantasize about it, we may hope for solar
intimacy to be non-possessive, non-hierarchical,
and non-individualized. Without the scarcity

The South Atlantic Quarterly 120:1, January 2021
DOI 10.1215/00382876-8795694 © 2021 Duke University Press

model of fossil fuel, maybe we could enter into some realm of being in which we can loosen our grip on things, on people, on the world that surrounds us. In that potential future, our subjectivities may be less bound by the systems of oppression, categorization, and anxious identification that we now experience and inhabit.

This future has the potential to be very bright. In solarpunk imaginaries and plans for off-grid communes, it has been already been imagined and partially enacted, as Rhys Williams (2019) and Elvia Wilk (2018) so cogently describe. When I first began to think about writing this essay, I planned to examine these countercultural instances of the potential rejection of the fossil-fuel regime. I read solarpunk stories and pored over intentional community websites, looking for the ways in which they subverted our current paradigms of inequality and power, not only in their energy sources but also in their relational structures. These seemingly utopian visions couldn't, however, keep my attention, in their liberal humanist assumptions that equality and non-hierarchical relationality could be obtained through earnestly willing it.

Instead, I was drawn to images that were not, at first glance, subversive or radical. Inscribed as they were in a popular culture version of self-care, they did not fit my idea of what a scholarly essay, no matter how informed by cultural studies it is, should examine. These came from what I quaintly imagine to be the other to my current academic life—Netflix watching, yoga practice, and French feminist texts that I haven't read since my zealous days as a deconstructionist undergraduate. Anglo (white, cis-woman) liberal popular culture would not seem, at first glance, like the best place to find new paradigms for a solar future. Nor, perhaps, would the French theory that was so prevalent in English departments in the 1980s. Why then is this jumble of images, ideas, and concepts connected in my mind to solidarity and to possible visions of solarity?

Best to start from the premise that I'm going to get any answer to that question wrong. And to carry on anyway. Because I do think there's something right in drawing upon what surrounds me, right now, right here, in the mainstream currents of my everyday life. In imagining the social and political possibilities that solar might offer, how else can I envision the future than to peer through the myopic lens of the present? The systems, structures, and objects that we have at our disposal *now* contain and conscript and expand our horizons. It's the conundrum of all utopian thinking, as Fredric Jameson (2004: 46) reminds us: "Its function lies not in helping us to imagine a better future but rather in demonstrating our utter incapacity to imagine such a future—our imprisonment in a non-utopian present without historicity or futurity—so as to reveal the ideological closure of the system in which we are somehow trapped and confined." Knowing this, I fear that every time I describe a possible opening of radical potential, my next sentence will reinscribe it within possessive individualism, body essentialism, or just plain sexism. I feel a little woozy already, just thinking about the sharp veering in and out of focus that I'm going to be enacting in this essay. Woozy from the woo-woo that we'll wander through together.

So, you see, this is how it goes for me as I try to write this paper. One moment I feel solarity to be full of liberatory potential, the next I suspect it will just be coopted into neoliberal entrepreneurialism. Back and forth between utopian future and non-utopian present. For each ray of sun there's a thick pall of smog.

Would I, however, know solarity when I saw it? Are there glimpses of it in my life right now? And can I celebrate those exuberant flashes of it, no matter from where they come? What tools do I have in my arsenal that enable me to imagine the unimaginable, the utopian horizon of an energy regime in which I (or someone yet to come) is a different kind of subject? That is, of course, the impossible—to conceptualize a subjectivity and a society that is not shaped by fossil fuels, even though everything we live and think and are is permeated by their dirty, polluting, viscous omnipresence. In what follows, I will share with you some of the strange associations that I have conjured as I attempt to imagine sol(id)arity.

Cultural critics and theorists who think about energy and the environment often turn to Georges Bataille's theory of the general economy. Stated most clearly in his *The Accursed Share* (written between 1946–49), this theory opposes capitalism's progress, in which accumulation and consumption occur within a restrictive economy, with a general economy in which the excess (of acquisition, of energy, of power) is expended, either through squandering, giving, exploding, or even going to war. Though he calls this share of the wealth "accursed," he believes that there is an ethics involved in giving it away, in surrendering it completely and without expectation of return. This is what was lost, he says in "The Notion of Expenditure," because of the bourgeoisie: "Everything that was generous, orgiastic and excessive has disappeared" (Bataille 1985: 124). In not spending the excess energy that is constitutive of life on earth, modern humans set themselves up for the horrors of war and other forms of mass destruction.

The example that is most central to his work is solar energy, which "is the source of life's exuberant development" (Bataille 1989: 28). In his attempts to explain the excesses of nature's relation to the sun, he describes a path, covered in asphalt, under which plant life "aspires in manifold ways to an impossible growth; it releases a steady flow of excess resources, possibly involving large squanderings of energy. The limit of growth being reached, life, without being in a closed container, at least enters into ebullition: Without exploding, its extreme exuberance pours out in a movement always bordering on explosion" (30). Thus, in Bataille's thinking, there is *too much* energy in the world, more than can even be consumed, accumulated, hoarded by the greediest of species. The tendency of this energy is to "pour out."

This image of extreme exuberance is not new to *The Accursed Share*, which is book-ended by other works in which he more explicitly explores expenditure and pouring out. Throughout his work, we find ejaculatory images: the volcano that erupts, the sun that directs "its ignoble shaft," the asshole that voids itself, the urine that flows. If this sounds eroticized, it's because it is. For Bataille, the erotic is a site where the accursed share can be wildly and freely spent. In early texts such as *The Solar Anus*, as well as in late texts such as *Erotism*, desire (both male and female) builds itself to inevitable climaxes, overflowing even the stringent containers put on it by the madhouse or the church or the other institutions that Bataille's characters and situations violate and defy.

In *The Story of the Eye*, a surreal text published in 1929, the narrator and his lover Simone have ceaseless orgasms, pissing and ejaculating in reactive response to round objects such as eggs, testicles, and eyeballs, and, importantly, the sun. Indeed, these objects, all repetitions of the eponymous eye, are the main characters of the story. These objects embody both roundness and fluidity, as Roland Barthes ([1963] 1982 :121) points out: "Is there anything more 'dry' than the sun? Yet, in the field of metaphor traced by Bataille. . . , the sun need only become a disc and then a globe for its light to flow like a liquid and join up, via the idea of a 'soft luminosity' or a 'urinary liquefaction of the sky', with the eye, egg, and testicle theme."

This, then, is Bataille's repeated refrain, that everything is constantly transforming, that what is known as one thing also contains within its opposite. Through such images as the solar anus (light and dark, day and night), he affords us "the image of an erotic movement that burglarizes the ideas contained in the mind, giving them the force of a scandalous eruption" (Bataille 1985: 8).

It is from these oppositions that the potential for revolution comes, for a radical transformation of the current order: "This eruptive force accumulates in those who are necessarily situated below. Communist workers appear to the bourgeois to be as ugly and dirty as hairy sexual organs, or lower parts: sooner or later, there will be a scandalous eruption in the course of which the asexual noble heads of the bourgeois will be chopped off" (8).

Lower parts—dirty, hairy, not idealized forms of beauty—become the solar image that transforms the revolutionary thoughts of the mind into a scandalous enactment. One thing necessarily transforms into another in the continuous circular movement that Bataille describes as his ethics of excess.

The movement between oppositions, between elevated concepts and the hidden orifices of bodies, is especially explicit in his novella, *Madame Edwarda*: "She was seated, she held one leg stuck up in the air, to open her crack yet wider she used fingers to draw the folds of skin apart. And so Madame Edwarda's 'old rag and ruin' loured at me, hairy and pink, just as full of life as some loathsome squid. . . . 'You can see for yourself,' she said, 'I'm GOD'" (Bataille 1997: 229).[2]

There is much that is hard to read in Bataille. The images can sometimes be revolting. Both men and women expose and exert their genitals in manic frenzies of masturbation and fucking. I try not to feel particularly perturbed by his depictions of female "rags" and "loathsome squid," because I get that it is in this very mixing of the profane and the sacred that his theory of a general economy lies. Yet, given that there has been a long history of the objectification of women's bodies, and given that, according to Peggy Orenstein (2016), it would seem that girls in our contemporary culture already feel that their vaginas are "ugly, rank, unappealing," I'm not sure that I can get behind a focus on vulvas as dark and repugnant. I would like to find possibilities in which women's "lower parts" can give us a figure for a different mode of relationality in ways that do not focus so much on sacred/profane as on shame/pride, or secrecy/openness.

My search for the utopian potential of Bataille's solar excess brings me, perhaps scandalously, to Gwyneth Paltrow. The rose vulva that I described at the beginning of this essay was a promotional set to inaugurate Gwyneth Paltrow's new television series, *The Goop Lab*. The show is the latest iteration of her Goop brand, which has, since 2008, given advice on style, beauty, food, health, and exercise, endorsed clothing brands and sex toys, and sold objects such as the infamous "This smells like my vagina" candle for $75. Yes, like Hadley Freeman (2020) and Arwa Mahdawi (2020), I too cringed, as I have when some of her bogus health claims have been critiqued by scientists and doctors.

What I am less eager to admit, however, is that I have a secret fondness for her. She is one of the few celebrities that I feel that I *know*, because she has lived so much of her life in public. I have cooked from her recipes, and I have followed her advice on exercise. I am heartened, therefore, by Elisa Albert and Jennifer Block's (2020) "Who's Afraid of Gwyneth Paltrow and Goop? The Long History of Hating on 'Woo'" because it calls out the sexism inherent in an outright rejection of Gwyneth's attempts at a lifestyle brand: "To be clear, we aren't looking to Goop for scientific rigor (or political consciousness, for that matter). But it's condescending to suggest that if we are interested in having agency over our bodies, if we are open to experiencing heightened states of awareness and emotion. . . , we are somehow privileged morons who deserve an intellectual (read: patriarchal) beat-down."

It is interesting that, though Albert and Block defend the contemporary (feminized) subject's right to explore alternative forms of knowledge, they still do not see *The Goop Lab* as a potential opening for political consciousness. I agree with them: it's not, not within the petroculture in which we now live. Gwyneth is thoroughly inscribed in and identified with neoliberal entrepreneurialism, speaking a hodgepodge of Silicon-Valley life-hack self-care jargon. This is most jarringly obvious in the opening sequence of the show, in which Gwyneth, sitting at the head of a board table, talks to her associates about the purpose of the show: "To me, it's all, like, laddering up to one thing, which is optimization of self. Like we're here one time, one life." This is the neoliberal subject par excellence, consistently seeking to extract highest possible value out of herself as human capital.

Gwyneth has, however, been a galvanizing force for one particular form of contemporary political consciousness: the #MeToo movement. Her speaking out about Harvey Weinstein was instrumental in the rape case brought against him, and opened the door for other women to come forward. In his "I Love Gwyneth Paltrow. There. I Said It" piece in the *New York Times*, Wesley Morris (2019) speculates about Gwyneth's fraught relationship to acting because of Weinstein's support of her. He goes through her filmography, and says that, even though she won the Oscar "basically for playing the sun" in *Shakespeare in Love*, she more often played "the dark parts," roles in which she was a female character who was sick, addicted, trapped, or in danger. He conjectures, "But maybe you get tired of all that darkness, of the suffering, of the being *made* to suffer. Maybe you really do just want the lightness of celebration, the benefit of salves and powders and whatever a jar of ashwagandha purports to do for you."

This juxtaposition of light and the dark seems to be the ways in which Gwyneth is figured, and figures her own understandings of the world. So much about her, her website, and her new show, is drenched in sunlight, reaching for a natural brightness that is warmly visible. As she has moved away from acting, she has made it her mission to bring light to the things that are often seen as dark. This is even the case in her own thoughts on Weinstein: "You know, I don't like to be binary about people or about things. I think we're all equal parts or varying percentages light and dark, and I think that, you know, he was a very, very important figure in my life" (Yaffe-Bellany 2019).

So when people started talking about *Goop Lab*'s Episode 3: "The Pleasure is Ours" (Paltrow 2020), I figured I should watch the show. In this episode, Gwyneth looks sun-kissed, with her blonde hair and "glow" makeup. Her dress is delicately covered with flowers and vines, as if they're growing over her. In front of her is a 90–year old woman who radiates energy from her clear eyes, her unadorned face. It is Betty Dodson, the famous sex educator who, thirty years ago, started running workshops in which women masturbated together. She is accompanied by her enthusiastic business partner, Carlin Ross. The lighting seems natural, as does the conversation, which illuminates the things that women keep hidden—that they are ashamed of the size of their labia, that they have never looked between their legs, that they don't know their desire. A particularly salient point is when Dodson explains to Gwyneth that the vagina is just the birth canal and the vulva is the whole thing (is this an opportunity for a candle rebranding?). The episode is, as many critics have said, educational.

As the women discuss labiaplasty surgery, which has increased worldwide because of women's genital dysmorphia (due, the show implies, to porn), Gwyneth says, "What we're talking about is this deep resistance we have to our own genitalia, culturally speaking. It is really time for that to change." Dodson and Ross describe what they do in their Bodysex workshops, in which women undress, show each other their vulvas, and then masturbate. Images of naked women talking in a circle are intercut. Gwyneth exclaims that there is no way that she would be able to do that. For a woman whose bikini photo shoots expose most of her assiduously-sculpted body, it is clear that she cannot part with those few inches of expensive fabric that hide her vulva from the camera.

Carlin, on the other hand, allows the camera to zoom in on her vulva, as she parts its lips and discusses its appearance with Dodson, who peers over one of her bent legs. As she says later, after masturbating herself to orgasm, "Somebody had to. You think of, like, feminism. Any time we took a step forward towards equality, it meant someone had to put it on the line." What, then, is put on the line in this show? Well, not just one vulva, but many, because, as Dodson says, "You can't show just one." The show sets out to illuminate its viewers about how different vulvas can be, showing a gallery of close-up images of varied shapes and sizes. They are anonymous, and rendered interesting, but not erotic.

I have never seen anything like that on mainstream television before. It feels simultaneously radical and dated to me. The feminist consciousness that was prevalent when I was a young woman, in which we were encouraged to own a speculum by collectives such as the authors of *Our Bodies, Ourselves*, seems to be making a necessary comeback. Of course, a critique can be levelled at this kind of liberal pro-sex attitude. In their description of the current-day continuation of "the orgasm politics of the 70s," C. E. (2012) could be responding directly to this *Goop* episode: "Using a rhetoric of personal agency, this sexual ethic of reclamation emphasizes the ability of the individual subject to attain a non-alienated state, not even through especially political means. All that is required is a lack of shame about sex and some control over how one wants to be fucked" (25). In my thinking on feminism up till now, I have, like C. E., critiqued the kinds of neoliberal rhetoric of empowerment that equate female agency with the ascent of individual women, celebrities like Gwyneth Paltrow. But I want to do something different here, because I'm looking, as I said, for those glimpses of solarity in amongst the consumer mainstream culture that shapes our ordinary affects and our cruel optimism.

So perhaps there is a different way to interpret the *Goop* vulvas. To see them not just as a mode of self-care in which the individual connects with her own subjectivity, but rather to see these vulvas as a challenge to the restrictive economy of the gaze, in which bits of flesh are parsimoniously exposed in order to elicit desire. What if, instead, we see this anonymous display of orifices, their dark interiors surrounded by illuminated labia, as a possibility, a counter-image to petro-masculinity? What if the vulva is to solar what the phallus is to fossil fuels?

From Bataille, I found myself turning to Luce Irigaray as I thought about vulvas. Having, in my 1989 variant of feminism, discarded her as an essentialist who equated female genitalia with gender, I decided to take another look. In her 1971 *This Sex Which Is Not One*, the lips figure both materially and symbolically, as a way of conceptualizing the otherness of the self to itself. This enables her powerful critique of the patriarchal phallogocentric masculine subject, which imagines itself to be self-same: "As opposed to the independent, autonomous individual of the phallic imaginary that has been bolstered throughout the history of philosophy, Irigaray shows that the interdependent and self-othering subject is figured by the image of the lips" (Anderson 2017: 64).

As has been made clear by many theorists of Irigaray as well as Irigaray herself, these lips are not biologically and fundamentally female, though they do function as part of the strategic essentialism that informs her work. As well, her category of "woman" is not to be understood as anything other than a social and historical construct that defines the other to phallogocentrism. Lynne Huffer (2011) has done important work to reconceptualize Irigaray's work as not just feminist, but powerfully queer: "As catachrestic heterotopias, the lips are both real and unreal, both what is and what is not. They cannot be pinned down as actual vaginas for the buttressing of cultural feminist projects: utterly unreal, they are neither here (on the mouth) nor there (between the legs)" (529). It is in this movement between the "here" and "there" that the lips have the potential to transcend binaries, to destabilize the masculine monosubject as well as the feminine cis-woman. In arguing for her place in queer theory, Huffer rearticulates Irigiray's work as profoundly non-normative in its insistence on alterity, on the ethics of difference.

Of course, I can only get so far reading *Goop Lab*'s vulvas through Irigaray's theories. So when Dodson speaks of the importance of women having orgasms, Gwyneth responds knowingly, "It's about you owning your own body . . . owning your own pleasure." This is where the potential of this radicality gets reinscribed into capitalist regime in which the pleasure is individual, owned by the individual. The vulvas may be anonymized, but ownership needs to be anxiously reasserted. This is a far cry from Irigaray (1985: 31): "Woman always remains several, but she is kept from dispersion because the other is already within her and is autoerotically familiar to her. Which is not to say that she appropriates the other for herself, that she reduces it to her own property. Ownership and property are doubtless quite foreign to the feminine."³ Within the version of feminist empowerment that Gwyneth has haphazardly assembled through different influences and currents, "owning your body" is the imperative and the rallying cry. In the depiction of the many vulvas, however, something more Irigarayian is happening, in which woman is several, in which a woman's otherness, which is a constitutive part of her, will never be possessed by her. There is no self-possession in this antihumanist philosophy.

There's no right or wrong to all of this. I'd like to just hold us all, Irigaray and Gwyneth and their supporters and detractors, in a (vulvar, solar) space in which we don't clamor to prove each other stupid, or mistaken, or blinded. Because we are all blind to what solarity could bring, and perhaps we can instead try to foreground the glimpses of it that shimmer in the background of our lives.

I wrote earlier about Gwyneth's opening sequence, in which she talks about optimizing the self. She continues, however, in a locution that veers from bro slang to obscenity: "Like, how can we, like, milk the shit out of this?" With this strange mixed metaphor, we could be back in Bataille's *Story of the Eye*, with its allusions to the fluids and excess that come from the orifices of the human body, to its pleasure in the erotic expenditure of bodily release. In this reading, we can see the conjoining of these corporeal emissions, that are neither "here" nor "there" (not literal milk and shit, not of the breasts and the anus) as yet another glimpse into a solarity that takes place in a space that does not yet exist, in bodies that will inhabit that space in ways that we cannot imagine. Gwyneth may not know what she is saying in this metaphor, but she obviously finds it compelling enough to include as part of the opening of each show in the season.

I want to end with something that I'm pretty sure both Gwyneth and Irigaray would like. It is the Divine Light mantra given to Sivanandha Radha (2010), a German-Canadian woman who went to India in 1956. I first learned it in yoga, when a member of our community tragically had a stillbirth the day before her baby was due. The senior teacher gathered us in a circle, and said that she had learned this mantra in the '70s. She told us that it was not religious, and that we could imagine "divine" as whatever felt right to us, whatever we imagined to be both bigger than us and simultaneously inside and outside of us. It at first felt "woo" to me, but I was so devastated that I chanted along with the others:

> *I am created by Divine Light*
> *I am sustained by Divine Light*
> *I am protected by Divine Light*
> *I am surrounded by Divine Light*
> *I am ever growing into Divine Light.*

Then our teacher asked us to imagine that a shower of brilliant white light was pouring down on us and filling our entire being. Once we were filled with it, we were to visualize it expanding outside of our own bodies, becoming a hovering sphere that incorporated our entire circle. "Now share the light with the grieving parents into this light, let it fill them and envelop them. Now put friends, family, anyone who you want to include. Now bring into this light the politicians and those people in power who have the task of making decisions and governing." It was certainly the first time I had positive thoughts towards then Canadian Prime Minister Stephen Harper.

In that collective invocation of light, I felt a way of being with others that was unfamiliar, and very welcome. All of us liberal, affluent, mostly white Anglos overcame our embarrassment and awkwardness at doing something that was so woo. We came together in a collective action that was not predicated on self-knowledge, or knowledge of the other, but rather on a shared giving over to the idea that there was something from which we could draw solace and sustenance. Divine light. The light of the sun. In that light, we cared for each other and ourselves in a way that was both thoroughly inscribed in a system of cultural appropriation and New Age spiritualism, and that also glimmered something that exuberantly exceeded that system. Or, as Irigaray (1992: 44) puts it: "Would you not want a solar flesh which was not fixed in the identity of a form?"

Notes

1   In *Hable con ella/Talk to Her,* Pedro Almodóvar (2002) incorporates a short silent movie, "The Shrinking Lover," in which a scientist takes a shrinking potion and slips past the giant labia into his lover's vagina. Allyson Mitchell's installation is described by curator Sarah Quinton as: "From floor to ceiling, folds of large pink pillows, soft drapes and carpeting comfort and cocoon. At the same time, the installation creates feelings of unease as the participant confronts their fears of claustrophobia, genitalia, mortality and appetites. . . . Pinker than pink, the Hungry Purse is a symbol of the opposing virtues of femininity and 'girl power'. It is a rumpus room and a vortex womb—an affirmation of the feminized body."

2   Quoted and brilliantly analyzed in Stoekl, 2007:106.

3   Quoted in Anderson 2017: 64–65.

References

Albert, Elisa and Block, Jennifer Block. 2020. "Who's Afraid of Gwyneth Paltrow and Goop? The Long History of Hating on 'Woo'." *New York Times.* Feb 3. nytimes.com/2020/02/03/opinion/goop-gwyneth-paltrow-netflix.html.

Almodóvar, Pedro. 2002. *Hable con ella/Talk to Her.* DVD. Sony Picture Classics.

Anderson, Ellie. 2017. "Autoeroticism: Rethinking Self-Love with Derrida and Irigaray." *PhoenEx* 12, no. 1: 53–70.

Barthes, Roland. (1963) 1982. "The Metaphor of the Eye," translated by J. A. Underwood, in Georges Bataille, *Story of the Eye,* translated by Joachim Neugroschel, 119–127. London: Penguin Books,

Bataille, Georges. 1985. "Solar Anus." *Visions of Excess: Selected Writings, 1927–1939,* translated by Allan Stoekl, 5–9. Minneapolis: University of Minnesota Press.

Bataille, Georges. 1989. *The Accursed Share, Vol. 1,* translated by Robert Hurley. Brooklyn: Zone Books.

Bataille, Georges. 1997. "Madame Edwarda." In *The Bataille Reader.* Edited by Fred Botting and Scott Wilson, 223–36. London: Blackwell.

C. E. 2012. "Undoing Sex: Against Sexual Optimism." *Lies: A Journal of Materialist Feminism.* 1: 15–44.

Freeman, Hadley. "Why is Gwyneth Paltrow Selling a Candle that Smells Like Her Vagina?" *The Guardian.* January 13. theguardian.com/fashion/2020/jan/13/why-is-gwyneth-paltrow-selling-a-candle-that-smells-like-her-vagina-goop.

Huffer, Lynne. 2011."Are the Lips a Grave?" *GLQ: A Journal of Lesbian and Gay Studies.* 17, no. 4: 517–42.

Irigaray, Luce. 1985. *This Sex Which is Not One*, translated by Catherine Porter with Carolyn Burke. Ithaca: Cornell University Press.

Irigaray, Luce. 1992. *Elemental Passions.* London: Routledge.

Jameson, Fredric. 2004. "The Politics of Utopia." *New Left Review*, no. 25: 35–54.

Mahdawi, Arwa. 2020 "Gwyneth Paltrow has Capitalized on Vaginal Shame and Celebration." *Guardian.* January 18. theguardian.com/world/2020/jan/18/gwyneth-paltrow-goop-capitalized-both-vagina-shame-celebration.

Mitchell, Allyson. 2004. "Hungry Purse: The Vagina Dentata in Late Capitalism." Textile Museum of Canada, Toronto, ON. http://www.allysonmitchell.com/project.html ?project=hungry-purse.

Morris, Wesley. 2019. "I Love Gwyneth Paltrow. There. I Said It." *New York Times.* September 30. nytimes.com/2019/09/30/movies/gwyneth-paltrow.html.

Orenstein, Peggy. 2016. *Girls & Sex: Navigating the Complex Landscape.* New York: Harper Collins. Kindle.

Paltrow, Gwyneth. "The Goop Lab: The Pleasure Is Ours." *Netflix* video. Jan 24. 36 mins. netflix.com/watch/81044717?trackId=14170289&tctx=0%2C2%2Cf7b159d3-e55b-4647-bcc8-8c27292fbcaf-1423496331%2C7c6383c7-10fe-48ff-84bd-7ca785bd81ee_50962044X3XX1583689715187%2C7c6383c7-10fe-48ff-84bd-7ca785bd81ee_ROOT.

Radha, Sivananda. 2010. *The Divine Light Invocation.* Kootenay Bay: Timeless Books. Kindle.

Wilk, Elvia. 2018. "Is Ornamenting Solar Panels a Crime?" *e-flux*, April 9. e-flux.com/architecture/positions/191258/is-ornamenting-solar-panels-a-crime/.

Williams, Rhys. 2019. "'This Shining Confluence of Magic and Technology': Solarpunk, Energy Imaginaries, and the Infrastructures of Solarity." *Open Library of Humanities* 5, no. 1, 1–35.

Yaffe-Bellany, David. 2019. "Gwyneth Paltrow on Goop and Beyond: 'It's a Process.'" *New York Times*, November 11. nytimes.com/2019/11/11/business/dealbook/gwyneth-paltrow-goop.html.

Nandita Badami

Let there Be Light (Or, In Defense of Darkness)

In the summer of 2015, a hundred and twelve families living in previously low- or un-electrified rural hamlets in Barabanki, a district in the North Indian state of Uttar Pradesh, opted to receive two lights and a mobile charging point in exchange for a flat monthly fee of Rs. 100 (a little less than USD 2). They were offered this service not by the Indian state, which had failed to provide adequate (if any) grid electricity to their homes, but by a private solar developer with startup grant funding from USAID, and awards from the National Geographic Society and the World Economic Forum.[1] Mera Gao Power (MGP)—meaning "electricity in my village"—set up a solar micro-grid in the area consisting of a collection of solar panels and a battery backup system, along with wires, bulbs, and charging points to deliver the electricity generated to each household that signed up for the program.

It is a common enough business model, with noble enough intentions. It works almost entirely on the presumption that the Indian state will continue to fail in its promise to deliver highly-subsidized electricity through the national grid to what is commonly referred to as "the last mile"— remote areas often overlooked by state services

The South Atlantic Quarterly 120:1, January 2021
DOI 10.1215/00382876-8795706   © 2021 Duke University Press

because of a lack of basic connectivity infrastructure, including, at times, paved roads. The immediate value proposition for MGP and other last mile micro-grid businesses comes from an attempt to divert funds from the families' black market kerosene budget towards solar power. There is much to be lauded about the business plan, but the concept is not without its drawbacks: the poorest in the country end up paying a much higher price for a basic need that is relatively affordable elsewhere.

It is not my intention, however, to critique the effort. Instead, I want to focus on the work of a group of political scientists that were attached to the project in order to measure the "social impact" of the pilot intervention. To do this, the political scientists administered three rounds of questionnaires to adopter families in the area. In the first round, conducted before the micro-grid was installed, respondents were asked a series of survey questions that covered a wide range of social issues, including household savings and the monthly kerosene budget ("How many rupees do you spend buying kerosene per month from the Public Distribution System?"; "How many rupees a month does your household save?"; "How many rupees a month do you spend?"), the frequency of sexual harassment incidents in the area ("How often are women subject to domestic violence in your hamlet?"; "Would you feel safer if there were more light?"; "Is there sufficient lighting for you to go outside during the dark?"), and some fairly leading questions about productivity ("What do you use lighting for? Studying? Do your children also use lighting? If yes, do they use it for studying?") (Aklin et al. 2017). The same questions were asked of the respondents six months after the panels had been installed and the two lights and charging point were up and running in each home, and once more about one year after the intervention was successfully off the ground.

The results, in short, were underwhelming. While the intervention had some minimal effects in increasing overall access to electrification, as well as in possibly effecting a slight decrease in expenditure on black market kerosene, nothing substantial changed by way of the other metrics of development the survey questions had attempted to gauge. As the research paper published a few months later lamented, there was "no systematic evidence for changes in savings, spending, business creation, time spent working or studying, or other broader indicators of socioeconomic development" anywhere to be found in the newly electrified Barabanki hamlets (Aklin et al. 2017: 1).

The study created ripples in the Delhi NGO circuit, and produced a series of scathing headlines, including "India's rural solar revolution hasn't delivered on its promise" (Gallucci 2017), "No evidence of socioeconomic benefits in India's rural communities after electrification: Researchers" (PTI 2017), "Solar microgrids insufficient to improve rural incomes" (Putty 2017), and an opinion piece, penned by one of the authors of the study, titled "Off-grid power not a silver bullet for rural India" (Aklin 2017).

Put in context, the scholars in question were in conversation with a set of studies that unabashedly celebrated the ability of interventions such as these to accomplish the very ambitious miracles they created their metrics to measure. *Their* intervention was designed to bring rigor to bear on such unabashed optimism. And yet—despite the validity of that epistemic battle—surely it shouldn't be news that two lights and a mobile charging point is no silver bullet?

Why does it seem reasonable to expect all these dreams of development and indicators of increased rationality to result from an intervention as limited as the installation of two lights and a charging point in a previously un-electrified home? The answer I develop here has to do with how our broad conceptions of rationality condition the assumptions that accompany the extension of light into areas that still experience large swaths of darkness at nighttime. I suggest that the epistemic consequence derives in part from the metaphors of light that are closely bound to our conceptions of rationality. These metaphors were born from and in turn inform a deep-seated notion of striving for "human betterment" as effected through that specific assemblage of human beings and material environments we call political economy.

While the work of the political scientists in question does not directly embody nor metonymically stand in for political economy in the classic sense of trade between nations and the laws that govern the production and distribution of wealth, randomized control trials of this variety are increasingly implicated in a systematic global flow of money, goods and ideas that some have called philanthrocapitalism: the phenomenon of philanthropic organizations using their funds to shape markets, particularly in developing countries (Economist 2006; Jenkins 2011). At a certain scale of wealth, it is no longer the practice simply to give money to enable human betterment (as we commonly understand the act of philanthropy, or even, to a certain extent, state-directed development). Rather, here, money is always accompanied by metrics to measure for impact and defined by precise indicators of betterment, often with an overt or implied expectation of cascading effects on the ability of humans to make ever-more rational economic decisions. If these effects do not materialize—as, for instance, when access to electric light fails to produce the requisite increase in study time for children in a far-flung Indian village—the system produces the aforementioned pithy public lamentations ("solar energy is no silver bullet"), and the money, methodologies, and metrics move on.

Yet, this particular example of solar electric light as the—sadly, failed—harbinger of increased rationality is an unusual one. It is not your run-of-the-mill philanthrocapitalist distribution of mosquito nets to the expectation of reduced rates of malaria infection. Its aim is as immediate, and yet, much more abstract. It wants to do more than simply eradicate disease. It wants to effect broadly agreed-upon socioeconomic indicators of progress: an increase in "savings, spending, business creation, time spent working or studying" (Aklin et al. 2017: 1). In other words, it is not enough simply to state that the installation of the solar micro-grid *worked* because it had the effect of increased access to electricity (a direct causal effect of the same variety as a decrease in rates of malaria infection).

My intervention is simply to ask: How did we get here? How did we arrive at our current and peculiar disposition of rationality, where access to light and electricity is not reason enough in itself to provide access to light and electricity? What makes it *rational* to expect these dreams of development and increased rationality to result from an increased access to light? And what bearing might this received rationality have on the epistemics that inform our transitions from fossil fuels to solar energies?

Metaphor, Metaphysics

To unpack this question, I mobilize what will seem like an unforgivably Eurocentric gesture doomed to appear a cringe-worthy caricature of social science inquiry: I harken back to the Greeks. This is because the Greeks—Socrates, Plato, and Aristotle, but also their contemporaries and followers—popularized a metaphysics of light that permeates the very structure of Western thought about thought itself, and the hegemonic traditions of rationality that followed. Plato's Cave, a situation of darkness and shadows that had to be forcibly left behind in favor of light and the world of "things," set the stage for discussions on the nature of understanding and reality. Philosophers have debated the significance of the cave in the metaphor, the role of human endeavor in finding a way out of it, and the epistemological privileging of vision that resulted from its imagery. All of these details have had a lasting influence on the structures through which we approach thought about thought. For the purposes of this article, I focus on the epistemological work performed by its deployment of light as a metaphor for knowledge.

Emerging out of the Cave, man—who had hitherto only seen shadows on the Cave walls, and mistaken those shadows for reality—was, for the first time, confronted by beings (and objects) illuminated by light. In Plato's allegory, the light is natural: it is from the sun, which doubles as The Idea of the Good. Both the Idea of the Good and the sun itself cannot be apprehended (nor metaphorically comprehended) directly. By extension, sunlight, too, has a complicated ontology: not quite a thing in itself, but knowable through that which it illuminates (Blumenberg [1957] 1993: 33). In order to fully understand this metaphor of light, however, we also need to appreciate that which it does *not* illuminate: the quality of darkness evoked as its foil. As Hans Blumenberg ([1957] 1993) explains in his foundational essay "Light as a Metaphor for Truth," the darkness of Plato's allegory is not the darkness of night, but the artificial darkness of the cave. For this reason, it is not accorded the circumstance of "natural," and does not exist in direct duality or opposition with daylight. While daylight is nature both enfolded into and actualized as understanding, "the brightness in which one moves, in which the world articulates itself, in which it becomes survey-able and understandable . . . and . . . thereby makes existence (*Dasein*) understandable to itself" (33); darkness here is "ontologically impotent," produced by "screening off" that which is naturally everywhere (32–33). Even the metaphorical darkness we encounter in Greek tragedy, Blumenberg explains, is "not . . . something that is to be sensed dimly," but rather must be "[elucidated] with a ruthlessly bright light" (33–34).

What we have in this somewhat originary structure of meaning is a metaphor that gives way to a metaphysics.[2] It is an abstraction that collapses the vision associated with light as "a way of expressing the naturalness of truth" into its opposite: transcendent light becomes the *condition* through which to conceptualize truth (33). From this time on, light and truth, ontologically intertwined, limit our ability to engage with darkness as truthful, meaningful, or progressive in its own right. Centuries later, we are still navigating a legacy of darkness overwhelmingly tainted with negative association, as that which is (and, imperatively, ought to be) vanquished by light (35–36).

Although the neutering of darkness in Plato was complicated by the Neoplatonics, who embraced the dualistic opposition of night and day (and thus accorded it a degree of the "natural" it was denied before), and by Cicero, who shifted the location of this transcendent light from that which is external to the human body to that which must be found internally within it (hence "insight," or internal sight), the negative bias persisted into Christianity. The Christian God, after all, gave to man light on the first day of creation. In doing so, he perpetuated the duality, casting darkness as synonymous with all that humanity ought to eradicate within itself: ignorance, crime, lust.

In tracing this transformation of light from metaphor to metaphysics in the Western philosophical tradition, we can see how light comes to inform not just how we conceive of truth and what is rational, but by extension, how thought about thought is itself implicated in this metaphysics to the point where we struggle to think about and knowledge and truth outside of these dualities.

Medium, Message

This received metaphysics of light—as the condition through which to know truth—continued into the modern age, made visible, at a glance, through the manner in which the Western pantheon chose to relate its histories back to itself: the Dark Ages through to the Renaissance, complete with broad narrative progress from darkness to light, culminating in the Enlightenment.

During the course of the Enlightenment, however, many of these received metaphysical structures were meaningfully complicated, if not completely inverted: as scientific development re-centered the human as a historical actor, light, too, was no longer conceptually cognate with divine revelation. Through reason and the scientific method, human beings were newly capable of revealing light and truth *for themselves* (McMahon 2018: 120). It was also a time when light became an almost-commonplace object of scientific inquiry: several scholars, most famously Isaac Newton, ran experiments to further develop their understanding of the theoretical properties of light. These studies resulted, famously, in major advances in the science of electricity as well as in the technologies for lamps and lighting.

It was lamp technology and candles more than electricity that had an immediate public impact on the politics of light during the Enlightenment. The long eighteenth century saw Europe's main cities attempt large-scale public illumination, or street lighting, for the first time. Historian Darrin McMahon (2018) makes the compelling case that "lighting made of the medium its message," that is, public lighting served "as a literal operationalization of the cultural and intellectual ideas of Enlightenment" (122). The very first attempts at public lighting, in Paris in 1667, took place concurrently with the creation a new urban police force. The two institutions—lighting and policing—were mutually complimentary and reinforcing. They dovetailed in a grand "civilizing" mission, "aimed to impose greater order, rationality and efficiency on urban life while combatting what were coming to be regarded as its 'pathologies,' such as crime, disorder, filth, disease" (127).

Public lighting on this scale had some predictable effects. Brilliantly lit European cities were elevated in the European imagination when compared to cities in Japan, China, and Egypt in terms that are now exceedingly familiar: "A lighted city was an enlightened city, thoroughly European, thoroughly modern" (136). European cities imagined themselves to be at the very forefront of progress, in part *because* the sudden mass access to light was equated with an increase in rationality. On a more symbolic register, it worked to undo Christian belief in evils that lurked at night. Combined with its association with the police, public light promoted notions of safety and created the conditions for "night life" as we would now recognize the phenomenon: shops and coffee houses, the bulwarks of both revelers and socially-minded thinkers, stayed open as late as 11:00 p.m. Whole new realms of experience and sociality, credited to the expansion of the conditions that birthed the Habermasian public sphere, opened up for the very first time (149).

Public illumination at this scale created a new register of darkness that fit neatly into the normative linear structure that defined much Enlightenment thinking. What was once simply darkness (nighttime, or the other of daylight), now signaled space that was *not* illuminated. Illumination thus "created a periphery beyond light," both in its immediate vicinity and across the globe: darkness was newly marked as that where literal and metaphorical illumination was not, or had not yet reached (147). The implications of this were felt not merely in the race politics of the colonial project but also in the intertwining epistemics of light with "growth," through the GDP, a calculative device first invented by Adam Smith during the Enlightenment.

It is here—in this historical moment—that we can recognize a new epistemological development, with light beginning to be conceptualized as a material *cause* of increased rationality in itself. It is in this moment that we can locate an answer to the question of *how it came to be* that light, rationality, progress, and human betterment or development as we know it today get refracted and extended beyond Europe, epistemically contained in fixed, particular relation.

Motive

As the above overview suggests, it is not just rational but *immanently rational* to expect increased rationality to result from an increased access to light, because our very conceptions of *what is rational* are so inextricably bound up in its metaphors and metaphysics. Today, the metaphorics of light have come full circle: light is now an economic dataset through which we gauge certain kinds of "truth." Take for instance an IMF working paper, released in 2019, titled "Illuminating Economic Growth." The paper provides a statistical framework to "illuminate the uncertainty in official GDP per capita measures . . . using satellite-recorded nighttime lights as an additional measurement of *true* GDP data" (emphasis mine) (Hu and Yao 2019). Or the publicly accessible World Bank (2010) geospatial luminosity dataset, which uses "nighttime lights satellite imagery [to] provide an alternative means for . . . estimates of *total* economic activity" (emphasis mine). A press release detailing an American National Science Foundation funded project similarly describes how, "For some, the amazing hues along Broadway, the Las Vegas Strip or the Sunset Strip in Hollywood mean a fun night out. For an economist, these dazzling lights signify people's pockets are flush with cash; and in fact a new study confirms it . . . nighttime lights, when seen from outer-space, reveal how economically developed an area is. . . . Researchers measure the glow irradiance of this artificial or 'human induced' lighting to study economic growth, poverty, health status, and environmental conditions. They use it to track GDP and other economic activity" (National Science Foundation 2011). In other words, the latest step in the epistemic transformation of light—which we have traced, so far, from metaphor to metaphysics, from medium to message—is its deployment as an *index* of rationality.

It is at this point that I would like to return to Barabanki, the political scientists and my motive for discussing the MGP solar micro-grid project in the opening section of this essay. Although the political scientists did not manage to measure any increase in rationality (however defined) as the direct result of increased access to light, the *fact* that they attempted to do so and failed, is significant. It is, ironically, why they hit on something important after all: in lamenting that an increase in access to light does not always result in an increase of so-called rational behavior, they not only proved the assumptions behind light-as-economic-dataset wrong but also revealed the absurdity of our inability to think of light as, well, simply light. The disappointment at the study's findings points to the cumulative *irrationality* of structures of meaning, notions of betterment, and means of distribution of wealth that cannot accept as "good" the good that comes from access to two lights and a mobile charging point, whatever that might be. If we cannot let two lights and a mobile charging point just be two lights and a mobile charging point, without having them bear the burden of solving everything from savings and study habits to sexual harassment, this is less an indication of the failure of the technology than it is of the failure of a rationality—

developed through the Enlightenment and seeped in the metaphorics of light—promoted by political economy.

The problem is simple: political economy, much like the GDP, operates as a universal and de-historicized "epistemic infrastructure" (Murphy 2017). It masquerades not as a historically specific conception of rationality, but as rationality itself. And, importantly, it offers few resources for thinking with darkness. One might argue, and rightly, that we do have traditions of darkness that are meaningful: darkness is the realm of poetry, of astronomy, and astrology. However, most of these registers have no place in the rationalities that sustain the systems of political economy. As critical race studies scholars have long argued, the disciplinary separation of the political from the economic has resulted in an erasure of attention to the politics of darkness from the mainstream operations and knowledge practices of capitalist economics (Darder 2011: 117–20). Ultimately, darkness, in the grand geospatialization of the GDP through nighttime luminosity datasets, functions merely as a foil for all the meaning associated with the different intensities of light. Darkness is still just backwardness in the rationality that rules the world.

Endarkened Solarities

While exploring the possibilities of the poetic in escaping the language of exchangeability, Franco "Bifo" Berardi mentions, in a throwaway sentence, the difficulties of "[breaking] free of the intellectual automatism of the dialectical happy ending" (Berardi 2012: 8). This is precisely the tension I find worth considering as we work to theorize the condition we are calling "solarities" in this special issue. Although the recalibration of our collective energetic infrastructural regimes from oil to solar might be read as a shift to a more sustainable and just future, we could invoke Berardi's comment as a caution. As other scholars have pointed out, solar energy, in particular, comes with an almost materially prepackaged ethics that we would do well to critically examine (Cross 2012). This ethics is often packaged and sold through conjuring utopian possibilities as we overcome the unjust obduracies of oil. Yet, particularly *as* we turn away from oil and the underground, it behooves us to be cautious of the epistemics we are stepping into, and of any mobilizations of light we use to express the relationalities we wish to see birthed from the of solar energies that will come to replace them.

The broadly agreed upon moral good today exists at the fine point of balance between the economic and the environmental. Projects like MGP's solar micro-grid are developed and funded in order that we may locate that elusive balance. But in the process, if, as we have seen, we lose sight of the conception of good *as something in itself,* our terms of engagement, our very frameworks of meaning need rethinking. Ultimately, the rhetoric of sustainable development and the balance between economy and environment it seeks still holds in place and builds on the legacies of political economy. Perhaps the question I am asking then is: what would it take to provincialize political economy? What would it take for light to *just be light,* and not the silver bullet of our political-economic dreams? By this I mean not just what it would take to allow for people to determine for themselves what should or shouldn't be considered productive in relation to the uses of light, but also to caution against academic tendencies that might turn light into a "theory machine" (Helmrich 2011).

To recover the epistemological possibility of knowing and counting in ways that confound the longstanding dualism of light and darkness in the Western tradition, particularly in relation to the political economies in which contemporary projects like MGP unfold, I suggest we consider solarities actively within an epistemics of *endarkenment.* To begin with, of course, such an epistemics would position itself against the instinctive indexing of darkness as a lack, as backwardness, idleness, or corruption. Conversely, such an epistemics would also enable relationalities and frameworks of meaning that simply sit with darkness as an end in itself, without collapsing it into a critique of productivity, such as might be the case in casting nighttime darkness as a space of regeneration and restoration that feeds into and subsidizes capitalist productivity. With such a register of darkness, as an end itself, we might, in time, feel our way towards a larger epistemics of ends in themselves, unencumbered by overwrought dualities. Such an epistemics might express rather than extinguish brightness, and allow, in time, for a framework of engagement with systems of political economic knowledge production that let the darkness in, so to speak.

Notes

The author thanks Bill Maurer, Valerie Olson, Melissa Wrapp, and the editors of this volume for their generous feedback and edits. She also thanks Selco Foundation and the Council on Energy, Environment, and Water for providing fieldwork access and assistance. This research was funded by grants from the Social Science Research Council and The Society for the Social History of Technology.

1   While MGP has traditional philanthrocapitalist roots, it went on to raise USD 2.5 million in equity funding through impact investment. The trades went through the Impact Investment Exchange (IIX), which describes itself as "the world's first social stock exchange" (IIX Global 2017).

2   Somewhat, because I have elided the so-called Pre-Socratics in this retelling.

References

Aklin, Michaël. 2017. "Off-Grid Power Not a Silver Bullet for Rural India." *SWI Swissinfo.Ch, Swiss Broadcasting Corporation* (blog), August 16. swissinfo.ch/eng/sci-tech/opinion _off-grid-power-not-a-silver-bullet-for-rural-india/43294158.

Aklin, Michaël, Patrick Bayer, S. Harish, and Johannes Urpelainen. 2017. "Does Basic Energy Access Generate Socioeconomic Benefits? A Field Experiment with Off-Grid Solar Power in India." *Science Advances* 3 (May): e1602153. doi.org/10.1126/sciadv.1602153.

Berardi, Franko "Bifo." 2012. *The Uprising: On Poetry and Finance.* Semiotext(e) Interventions Series 14. Cambridge: MIT Press.

Blumenberg, Hans. (1957) 1993. "Light as a Metaphor for Truth: At the Preliminary Stage of Philosophical Concept Formation." In *Modernity and the Hegemony of Vision*, translated by Joel Anderson. Berkeley: University of California Press.

Cross, Jamie. 2019. "The Solar Good: Energy Ethics in Poor Markets." *Journal of the Royal Anthropological Institute* 25, no. S1, 47–66.

Darder, Antonia. 2011. "Chapter 6: What's So Critical about Critical Race Theory?: A Conceptual Interrogation With Rodolfo Torres." *Counterpoints* 418: 109–29.

Economist. 2006. "The Birth of Philanthrocapitalism." *Economist*, Special Report, February 25. economist.com/special-report/2006/02/25/the-birth-of-philanthrocapitalism.

Gallucci, Maria. 2017. "India's Rural Solar Revolution Hasn't Delivered on Its Promise." *Mashable*, May 17. mashable.com/2017/05/17/india-rural-solar-revolution-results/.

Helmrich, Stephen. 2011. "Nature/Culture/Seawater." *American Anthropologist* 113: 132–44.

Hu, Yingyao, and Jiaxiong Yao. 2019. "Illuminating Economic Growth." IMF Working Papers. April 9. imf.org/en/Publications/WP/Issues/2019/04/09/Illuminating-Economic-Growth-46670.

IIX Global. 2017. "Who We Are." iixglobal.com/who-we-are/.

Jenkins, Garry W. 2011. "Who's Afraid of Philanthrocapitalism." *Case Western Reserve Law Review* 61, no. 3: 786–87.

McMahon, Darrin M. 2018. "Illuminating the Enlightenment: Public Lighting Practices in the Siècle Des Lumières." *Past and Present* 240, no. 1: 119–59.

Murphy, Michelle. 2017. *The Economization of Life.* Durham, NC: Duke University Press.

National Science Foundation. 2011. "Nighttime Lights Clarify Economic Activity." *nsf.gov*, June 10. nsf.gov/discoveries/disc_summ.jsp?cntn_id=119737.

PTI. 2017. "No Evidence of Socioeconomic Benefits in India's Rural Communities after Electrification: Researchers." *Financial Express*, May 18. financialexpress.com/economy/no-evidence-of-socioeconomic-benefits-in-indias-rural-communities-after-electrification-researchers/675285/.

Putty, Madhukara. 2017. "Solar Microgrids Insufficient to Improve Rural Incomes." *SciDev. Net Asia and Pacific Edition*, August 6. scidev.net/index.cfm?originalUrl=/asia-pacific /energy/news/solar-microgrids-insufficient-improve-rural-income-alternative-power -energy.html&.

Smith, Adam. 1776. *The Wealth of Nations.* Modern Library.

World Bank. 2010. "Gross Domestic Product (GDP) from Night Lights." World Bank Data Catalog. datacatalog.worldbank.org/dataset/gross-domestic-product-gdp-night-lights-2010.

Jordan B. Kinder

Solar Infrastructure as Media of Resistance, or, Indigenous Solarities against Settler Colonialism

The ongoing history of setter colonialism, including in Canada, is inextricable from the infrastructures of energy and extraction that provide its material foundation. Addressing this relationship, this article explores how Indigenous solarities resist extractivism and generate conditions for just energy futures beyond settler colonialism through emergent solar infrastructures.

In a four-part article titled "Solar Panels and Sisterhood" published in Canada's *National Observer*, Michif nomad journalist Emilee Gilpin (2017) provides an account of the efforts of the Secwepemc-led Tiny House Warriors movement. The animating objective of this movement is self-determination: to occupy unceded, unsurrendered territory, while providing housing to those most affected by the persistence of settler colonialism, particularly women and children. The Tiny House Warriors are engaged in a project to construct mobile tiny houses along the path of the federally owned Trans Mountain Expansion Pipeline Project (TMX) as a means of providing possibilities for life against the necropolitical forces of what Andreas Malm (2016), Bob Johnson (2019), and others name the fossil economy. Gilpin details the partnership between Lubicon Cree energy justice

The South Atlantic Quarterly 120:1, January 2021
DOI 10.1215/00382876-8795718   © 2021 Duke University Press

activist Melina Laboucan-Massimo and the Tiny House Warriors to install mobile solar infrastructure for the community. The built environment here—tiny houses, solar panels—is leveraged against the encroachment of extractivist, settler colonial infrastructures.

This opening précis makes two important points regarding infrastructural inequality that serve as thematic bookends for what follows. First, it makes visible the nexus of extractivist infrastructural development and settler colonialism as a persistent formation. Second, it identifies collaboration between Indigenous actors from different nations who share resources and knowledge in gestures of deep solidarity against the mechanisms of settler colonialism. Laboucan-Massimo (2020) describes the partnership in terms of a solidarity stemming from "the same concerns about violence against Indigenous women especially when the rates of violence are exacerbated in resource extraction zones." When informed by energy justice, solarity is a form of solidarity among the human and nonhuman world; it describes a relation towards the sun that reorients our collective energy imaginaries from one of scarcity to one of abundance. Indigenous solarities, then, name the myriad potentialities for doing so in ways that foreground Indigenous self-determination against and beyond extractivism.

In approaching Indigenous solarities as grounded in material efforts to resist the inertia of the settler colonial, fossil economy, I explore here the limits and possibilities of emergent solarities in working towards a just energy future. Solarity in the first instance promises a *potentiality* for returning to an energy regime based on flows—wind, water, sunlight—from an energy system based on stores or stocks, a determinant condition of the fossil economy (Malm 2016: 38–42). However, while flows tend towards a commons (Malm 2016: 373), they do not inherently promise or produce good social and ecological relations. A solarity tethered to the solar panel as its primary fuel-gathering mechanism is in many ways an ambivalent condition. Its materialization could produce colonizing projects indistinguishable from its extractivist counterparts, a possibility seen in the failed DESERTEC solar farm in Northern Africa (Batel and Devine-Wright 2017: 6–7). Solar panels themselves have material requirements tied to extraction that have been framed in terms of "green extractivism" (Riofrancos 2019). Such contradictions underwrite struggles for a more equitable future, but they need not foreclose efforts to work towards a just energy future. Understanding the material-infrastructural modes of transition as emergent and dynamic rather than fixed and totalizing creates spaces of opportunity, experimentation, and disruption that are necessary for just energy transitions.

Developing a preliminary theory of Indigenous solarities, I anchor my observations to Sacred Earth Solar's two completed projects. I look to the Piitapan Solar Project in Laboucan-Massimo's home community of Little Buffalo, Alberta, Canada, and to the partnership with the Tiny House Warriors described above. My approach is methodologically informed by recent infrastructural thinking from theorists such as Lauren Berlant (2016) and Deborah Cowen (2018, 2019), both of whom offer expansive, relational understandings of infrastructure against dominant technological ones. It is also informed by thinkers such as Myles Lennon (2017) and Dagmar Lorenz-Meyer (2017) who respectively see in solar energy infrastructures the possibilities to decolonize energy and to generate a feminist techno-ecological ethos.

The Inertia of Extractivism:
Energy Infrastructure and Settler Colonialism Today

In settler colonial nations, widespread resource extraction and infrastructural developments for the primary benefit of transnational capital reproduce settler colonial relations. The day-to-day smooth operation of systems that undergird modernity rely on this continued dispossession, and those who live under the dominant energy regime are compelled to participate in these networks of exploitation through conscious and unconscious participation in petroculture (LeMenager 2014; Wilson, Carlson, and Szeman 2017). The reproduction of settler colonial relations is especially apparent in infrastructural developments tied to energy, including hydroelectric megadams, oil and gas pipelines, coalmines, fracking wells, and uranium mines linked to nuclear energy production and waste storage (Runyan 2018). Each new project adds another layer of intensification to the growing stack of integrated extractivist energy systems. The consequence of this intensification is an infrastructural inertia that materially binds the future to the conditions of the present (Malm 2016: 9). Against the inertia generated by these infrastructures deemed "critical" by the state, Indigenous resistance is prevalent precisely because critical infrastructures represent some of the most concentrated sites that reproduce settler colonial relations (Pasternak and Dafnos 2018; Cowen 2018; Spice 2018).

From the Sioux-led struggle against the Dakota Access Pipeline at Standing Rock to the Unis'tot'en camp's permanent occupation of traditional territory to block infrastructural encroachment, struggles over energy infrastructures and the futures they forecast are at the center of resistance to the continuation of settler colonialism. As I write this (February 2020),

Wet'suwet'en land defenders and their supporters at the Unis'tot'en camp blocking TC Energy Corporation's Coastal GasLink Liquid Natural Gas Pipeline have experienced raids and arrests from militarized police. Part of the first actions by the Royal Canadian Mounted Police (RCMP) included setting up an immediate exclusion zone for media and outsiders (MacLeod 2020). By starving access in this way, the RCMP took control of the flows of goods and information in order to isolate land defenders and to contain their sovereignty both materially and culturally. In response, Indigenous and non-Indigenous supporters have, among other actions, staged a series of railway, bridge, and highway blockades across Canada, strategically disrupting infrastructural choke points to slow the day-to-day circulation of goods, people, and, capital. These actions have had immediately measurable impacts on the economic landscape of Canada, impacts that Canada National Railway's CEO J. J. Ruest described as "significant" (qtd. in Jackson 2020).

Struggles over, against, and through infrastructure underwrite Canada's history up to the present. That railways are primary targets for actions in support of the demands of Wet'suwet'en land defenders is apropos given that railways are a metonymic expression of nationalism in the popular imaginary and hold continued political economic significance. In "The Jurisdiction of Infrastructure: Circulation and Canadian Settler Colonialism," Cowen (2018) argues that nation-binding infrastructures such as the Canadian Pacific Railroad initiated a self-fulfilling process of settler claims to jurisdiction over land and resources. As Cowen (2018) writes, "historical struggles over rail and contemporary struggles over pipelines both suggest that infrastructure and jurisdiction are deeply entangled in the making of settler colonialism." By establishing corridors for the flow of commodities in these ways, the Canadian state constructed infrastructural assemblages above and beneath so-called Crown Land that materialized this jurisdictional claim. According to Cowen, the Tiny House Warriors resist both the Trans Mountain Expansion Pipeline Project's encroachment on and implicit claims to unceded Secwepemc territory, and the legacies of settler colonialism propelled by such infrastructural development. Pipelines may have displaced the railway as the *prima facie* infrastructure that binds the Canadian national imaginary (Barney 2017: 80), but the railway remains a persistent medium of struggle through which Indigenous peoples express sovereignty by disrupting the circulation of capital (Pasternak and Dafnos 2017).

This materialization of settler colonial jurisdictional claims, claims that produce a settler colonial infrastructural *inertia*, is partly animated by what literary and cultural theorist Jennifer Wenzel (2020) calls "resource logics." Resource logics are "habits of mind through which humans understand nature as something other than themselves, disposed for their use (as a 'resource'), and subject to their control." In the deployment of such logics "nature has always-already entered economics, as 'natural resources': stuff-waiting-to-be-sold-and-used, along the lines of what Heidegger called *das Bestand*, the standing-reserve" (148). The kinds of infrastructures that support these logics are those that serve capital accumulation in the first instance—deemed critical infrastructures by the Canadian state—and comprise what Tlingit member of Kwanlin Dun First Nation and anthropologist Anne Spice (2018) understands as "infrastructures of invasion" (45). If the imaginaries that animate the fossil economy and its invasive infrastructures are tied to resource logics, then it is through a more expansive understanding of infrastructure that alternative imaginaries might emerge. Indigenous solarities offer an orientation that materially embody and express this expanded sense of infrastructure. A return to harnessing flows in ways that foreground Indigenous epistemologies and ways of being disrupts resource logics and their operations. In other words, Indigenous solarities *unsettle* extractivism because the infrastructure they develop to mediate energetic relations are premised on maintaining good relations between human and more-than-human kin (Spice 2018).

Indigenous Solarities as Media of Resistance: Enabling Self-Determination

Infrastructures of Indigenous solarities are media of resistance. Media theorist John Durham Peters's (2015) call for infrastructuralism as a way of orienting critical perceptions to the relationship between social, built, and natural environments offers a vocabulary for approaching the infrastructures of Indigenous solarities. Following Peters's articulation of media as "*enabling environments*" (46), the infrastructures of Indigenous solarities are media that enable environments of self-determination through what Dene theorist Glen Coulthard (2014) calls "grounded normativity." Grounded normativity describes "modalities of Indigenous land-connected practices and long-standing experiential knowledge that inform and structure our ethical engagements with the world and our relationships with human and nonhuman others over time" (Coulthard 2014: 12). The dominant energetic and infrastructural relations of today—those wedded to the fossil economy and extractivism—have served alongside a settler colonial state apparatus to hinder such land-connected practices and to undermine Indigenous sovereignty. Indigenous solarities resist these undermining tendencies.

As media of resistance, the infrastructures of Indigenous solarities refuse the extractivist conditions of the present while prefiguring desirable conditions for the future. In a collection of essays on recent protest camps, editors Gavin Brown, Anna Feigenbaum, Fabian Frenzel, and Patrick McCurdy (2017) develop the concept of *media of resistance* to examine how these camps strategically produce, engage, and mobilize media. To communicate internally and externally, camps both rely on existing and spontaneously generated modes of communication. Media of resistance enable a stronger internal cohesion and organization in tenuous environments. Equally important is how they amplify the messages at the core of the encampments and offer the potential for activating a broader coalition of support—crucially all done on an encampment's own terms (Brown et al. 2017: 12–13). An expansive definition of media—Peters's *enabling environments*—implies a similar broadened scope for media of resistance. What I propose here is that understanding solar energy infrastructures as media brings the cultural and material relations they mediate into sharper focus.

Sacred Earth Solar explicitly articulates the cultural and political narratives that these initiatives embody and express. Its first installation, the Piitapan Solar Project, was completed in 2015. The 20.8 KW/h installation powers the Little Buffalo community health center and was intentionally constructed with involvement from the local Lubicon Cree community. Its design and construction accounts for its social and cultural role in the community beyond merely fueling the community health center. The eighty raised racking panels have a minimized footprint—they offer a safe environment that does not need to be walled off and are, instead, integrated into the landscape. *Piitapan*, as Laboucan-Massimo describes in the first episode of *Power to the People*, her documentary series on Aboriginal Peoples Television Network (APTN), is a Cree word meaning a new dawn, which serves to signal transition and express self-determination beyond the confines of the fossil economy.

Transition is a keyword underwriting Sacred Earth Solar's broader mission. Piitapan itself is an attempt to work towards a more just energy future that directly confronts the social and ecological destruction endemic to fossil capitalism. It is an effort to establish alternative energetic and infrastructural relations and imaginaries in a community deeply impacted by an extractivist energy regime neither of its desire nor its making—what Laboucan-Massimo calls a "foreign imposition" (Power to the People 2020). Laboucan-Massimo's motivation for the Piitapan project came out of her community's direct experiences of the devastation caused by a major oil spill thirty kilometers from Little Buffalo in 2011. This was the second-largest oil spill in Alberta's history, with 4.5 million liters (nearly 1.2 million gallons) leaking into beaver ponds and muskeg from a Plains Midstream Canada-owned pipeline (CBC News 2013). Fumes from the spill affected the community, resulting in school closure. As a response to this trauma, Piitapan ultimately signifies "energy transition, self-determination, food security, and what autonomous decision making looks like" (Power to the People 2020).

By partnering with the Tiny House Warriors, Sacred Earth Solar has expanded its loci of resistance from responding to past ecological devastation to confronting the continued extractivist futures that the TMX embodies. Since infrastructural developments like pipelines must operate for decades before they become profitable, the TMX represents a commitment to an extractivist, fossil fuel future that is premised on the continued disposses-sion of Indigenous peoples. At the center of the Tiny House Warriors reoccu-pation of their traditional territory, and the project of Indigenous solarities more generally, is a refusal of this perpetually carbon-locked future.

Indigenous Solarities as Indigenous Feminism: Fueling Social Reproduction

As media of resistance, the infrastructures of Indigenous solarities express principles of Indigenous feminism. In "Badass Indigenous Women Care-take Relations: #StandingRock, #IdleNoMore, #BlackLivesMatter," Dakota feminist scholar Kim TallBear (2019) reflects on the crucial role of women at the forefront of social and ecological justice movements. Importantly, Tall-Bear (2019) "see[s] all of these women caretaking their peoples and kin, and [she watches] closely the conceptual, emotional, and organizing work their movements for change take" (16). Indigenous solarities are generated out of and in support of these conditions and movements for change. As media, infrastructures are more than their cumulative material functions and effects, and the infrastructures of Indigenous solarities are an expression of an Indigenous feminism that centers the maintenance of good relations between human and more-than-human kin.

The maintenance of good relations in this way can be understood in terms of social reproduction. Marxist feminists theorize social reproduction in terms of gendered, feminized, and unpaid labor that capitalism excludes from traditional structures of valuation yet, ultimately, rests upon for its con-tinued existence (De'Ath 2018). A broader view of social reproduction in this way is a significant vector through which to conceptualize the stakes and interventions of Indigenous solarities, since the infrastructural impetus of

Indigenous solarities resists destructive social and ecological practices whose effects are also highly gendered. Rebecca Jane Hall's (2016) work on reproduction and resistance in the context of anti-colonialism mobilizes social-reproduction feminism to theorize how Indigenous women's productive and reproductive labor challenges masculinist, capitalist modes of production by working outside of those economic structures. "Indigenous women resist the totality and violence of capital through multiple strategies," Hall (2016) explains, "whether it is through the persistence of non-capitalist Indigenous labor, through the privileging of reproductive work and interpersonal relations over the demands of capital, or both" (105).

An expanded notion of social reproduction in these terms illuminates the relationship between land and the more-than-human world centered by Indigenous solarities. In *As We Have Always Done: Indigenous Freedom through Radical Resistance*, Michi Saagiig Nishnaabeg theorist Leanne Betasamosake Simpson (2017) describes land-based pedagogies as a method through which Nishnaabeg intelligence is taught and practiced as theory (151). Simpson articulates how "the source of our knowledge" (2017: 172), a form of social reproduction, is tied to land. These land-based practices, knowledges, and theories are categorically disrupted and threatened through extractive processes and infrastructures in ways inextricably linked to gendered forms violence.

Efforts from Sacred Earth Solar are explicitly positioned against the gendered violence that underwrites extractivism. Gendered violence against Indigenous women, girls, and 2SLGBTQQIA persons is endemic to territories that extractive industries operate. The final report of Canada's National Inquiry into Missing and Murdered Indigenous Women and Girls (2019: 584), for instance, heard from "Expert Witnesses, institutional witnesses, and Knowledge Keepers . . . that resource extraction projects can drive violence against Indigenous women in several ways." Such violence is ultimately an expression of what T. J. Demos (2018) presciently calls the "necropolitics of extraction." As Laboucan-Massimo (2017) writes in her contribution to *Whose Land is it Anyway? A Manual for Decolonization*, "Violence against the earth begets violence against women" (39).

To fuel social reproduction through good energetic relations is to recognize the determinant role that infrastructure plays in shaping social relations. Laboucan-Massimo's own experiences, inspirations, and commitments as expressed through Sacred Earth Solar speak to this relationship between life, death, and infrastructure, particularly in terms of the ways that extractivist infrastructures threaten the lives of Indigenous communities. As Laboucan-Massimo explains in *Power to the People*, Piitapan's role is in serving an inter-generational community, offering a glimpse of a future not simply in terms of a transition to renewable energy sources, but in terms of livelihood. It is also not a coincidence that a key figure in the Tiny House Warriors movement, Kanhaus Manuel, is also a birthkeeper trained by Mopan Mayan midwives (Berman 2016). For Laboucan-Massimo, Manuel, and the Tiny House Warriors, the survival of their peoples and cultures are threatened by extractive energy regimes and Indigenous solarities informed by Indigenous feminism resists this thread and creates the conditions for resisting extractivism and establishing decolonized energetic relations on their own terms.

Indigenous Solarities as Indigenous Futurisms: Prefiguring Just Energy Futures

The infrastructures of Indigenous solarities are also infrastructures of Indigenous futurisms. Most closely associated with speculative and science fiction-based Indigenous cultural production, Indigenous futurisms describe a genre of work that articulates Indigenous visions of the future. Against a settler colonial imagination that categorically envisions a future free from Indigenous peoples, Indigenous futurisms imagine futures in which Indigenous peoples are, instead, both centered and central (Dillon 2012, 2016). Linking artistic imaginings of Indigenous futures to interventions upon the built environment in this way further underscores the necessity of thinking infrastructures as media, as enabling environments for Indigenous futures.

Infrastructures are material promises for the future (Anand, Gupta, and Appel 2018). This argument been taken up primarily to critique those infrastructures that continue to deepen dominant relations, including those tied to the nexus of extractivism, the fossil economy, and settler colonialism. But these promises manifest not only in terms of futures that deepen the dominant relations of the present; infrastructures can also represent a material promise for a future *beyond* these relations. The material promise for the future that Indigenous solarities offer is one of possibility in this way. Actually existing materializations of Indigenous solarities today mobilize photovoltaic technologies as a speculative gesture of resistance and *potential*, challenging the totalizing tendencies of the fossil economy by interrupting and displacing its stronghold.

Indigenous solarities are a kind of futurism that fuse solar technologies with Indigenous epistemologies and ways of being to promise a decolonial, just energy future. In contemporary architecture and design theory that addresses how we might build a climate resilient future, Indigenous design traditions have emerged as a focal point. In *Lo-TEK. Design by Radical Indigenism*, designer and environmentalist Julia Watson (2019) proposes the notion of Lo-TEK to describe Indigenous design traditions and technologies premised upon ways of being with nature rather than against it. As a portmanteau that combines lo-tech and Traditional Ecological Knowledge (TEK), "Lo-TEK counters the idea that indigenous innovation is primitive and exists in isolation from technology" (21). As a design movement, Lo-TEK seeks to "rebuild an understanding of indigenous philosophy and vernacular architecture that generates sustainable, climate resilient infrastructures" (21). The aim of Watson's compendium of Indigenous design traditions and technologies, which is to "[rewrite] the mythology of technology" (21), is needed, but there arguably remains a residual fetishization of the *lo* that narrowly defines Indigenous technological relations.

Indigenous solarities like those pioneered by Laboucan-Massimo's Sacred Earth Solar do not fit cleanly within these particular parameters of Indigenous innovation. Yet they are an expression of the very principles that underwrite Lo-TEK's philosophy of innovation and sustainability. In Lo-TEK's vision it is not entirely clear for and by whom these Indigenous futures are made: it privileges the contributions that Indigenous technologies can make to a broader design program, but in many ways avoids naming the settler colonial practices and relations that have marginalized Indigenous technologies in the first place, opting instead to identify ideologies and mythologies of Enlightenment as the culprit. A lesson can be learned from Laboucan-Massimo's *Hi-TEK* vision for a sustainable energy future: that a just energy future is one that privileges the self-determination and sovereignty of Indigenous nations and peoples.

Conclusion: Energy Futures between Foreclosure and Possibility

Gilpin (2017) pointedly begins her article on the partnership between Laboucan-Massimo and the Tiny House Warriors by declaring that its narrative impetus "is not a protest story." Too often, Indigenous resistance to settler colonialism is construed in the popular imaginary as singular-cause protest. Writing in the wake of the NoDAPL movement at Standing Rock, Lakota theorist and historian Nick Estes (2019) echoes Gilpin by highlighting how "the protestors called themselves Water Protectors because they weren't simply against a pipeline; they also stood for something greater: the continuation of life on a planet ravaged by capitalism" (15). Alternative vocabularies from the frontlines resist externally imposed definitions of the motivations, demands, and desires of particular actions of resistance. In the contemporary realities of ongoing settler colonialism, what is perceived as a single-issue is tied to a broader set of interlinking relations. Estes (2019) closes *Our History is the Future* with a meditation on the solidarity that stems from shared material struggles. "New pipelines are creeping across the continent like a spiderweb, with frightening speed, but in the process, they are also connecting and inciting to action disparate communities of the exploited and dispossessed," he writes (253).

When the histories and presents of energy and infrastructure are bound to continued legacies of settler colonialism, energy futures that hope to be truly equitable must break from reproducing these legacies. As energy humanities scholar Sheena Wilson (2017) reminds us, "Energy transition politics can continue to intensify inequities grounded in particular epistemologies, or introduce new alternatives" (404). In this article, I have focused on two solar infrastructure installations by Sacred Earth Solar as examples of these alternatives. Yet, it is important to underscore that the motivations for Indigenous renewable energy projects are as multiple and complex as the belief systems and practices that undergird them. These preliminary approaches to Indigenous solarities emerge from specific contexts; while they may travel well to other locales and regions, they are by their nature situated. Across Canada, however, Indigenous, community-based renewable energy initiatives have emerged as a means to materially assert energy autonomy (Stefanelli et al. 2019). Across the planet, Indigenous, community-based renewable energy projects belong to a larger paradigm of energy democracy (Fairchild and Weinrub 2017). Solarities do not simply name a *technological* relation to the sun, but rather, a broader social relation reoriented from stock back to flow, and these interlinked renewable energy infrastructure initiatives mutually inform each other. Through this expanded, relational understanding of energy infrastructure, Indigenous solarities mediate relations and modulate energy futures between extractive foreclosure and post-extractive possibility.

Note

This article owes a great deal to my participation in the *Imagining Indigenous Solarities: 7th Generation Character Design Workshop,* led by Maize Longboat and Waylon Wilson as part of *Solarity: After Oil School II.* I thank Maize and Waylon, along with Darin Barney, Emilee Gilpin, Derek Gladwin, Melina Laboucan-Massimo, Elizabeth Miller, Simon Orpana, Shirley Roburn, and Kyle Whyte, whose generous and insightful collective discussion helped refine much of what appears here.

References

Anand, Nikhil, Akhil Gupta, and Hannah Appel, eds. 2018. *The Promise of Infrastructure.* Durham, NC: Duke University Press.

Barney, Darin. 2017. "Who We Are and What We Do: Canada as a Pipeline Nation." In *Petrocultures: Oil, Politics, Culture,* edited by Sheena Wilson, Adam Carlson, and Imre Szeman, 78–119. Montreal, QC: McGill-Queen's University Press.

Batel, Susana, and Patrick Devine Wright. 2017. "Energy Colonialism and the Role of the Global in Local Responses to New Energy Infrastructures in the UK: A Critical and Exploratory Empirical Analysis." *Antipode* 49, no. 1: 3–22.

Berlant, Lauren. 2016. "The Commons: Infrastructures for Troubling Times." *Environment and Planning D: Society and Space* 34, no. 3: 393–419.

Berman, Sarah. 2016. "An Indigenous Mom Explains Why She Doesn't Register Her Kids with the Government." *Vice,* June 26. vice.com/en_ca/article/7bmp7x/an-indigenous-mom-explains-why-she-doesnt-register-her-kids-with-the-government.

Brown, Gavin, Anna Feigenbaum, Fabian Frenzel, and Patrick McCurdy. 2017. "Introduction: Past Tents, Present Tents: On the Importance of Studying Protest Camps." In *Protest Camps in International Context: Spaces, Infrastructures and Media of Resistance,* edited by Gavin Brown, Anna Feigenbaum, Fabian Frenzel, and Patrick McCurdy, 1–22. Bristol: Bristol University Press.

CBC News. 2013. "2nd Largest Pipeline Spill in Alberta History Leads to Charges." *CBC,* April 26. cbc.ca/news/canada/edmonton/2nd-largest-pipeline-spill-in-alberta-history-leads-to-charges-1.1311723.

Coulthard, Glen. 2014. *Red Skin, White Masks : Rejecting the Colonial Politics of Recognition.* Minneapolis, MN: University of Minnesota Press.

Cowen, Deborah. 2018. "The Jurisdiction of Infrastructure: Circulation and Canadian Settler Colonialism." *Funambulist* 17. thefunambulist.net/articles/jurisdiction-infrastructure-circulation-canadian-settler-colonialism-deborah-cowen.

Cowen, Deborah. 2019. "Following the Infrastructures of Empire: Notes on Cities, Settler Colonialism, and Method." *Urban Geography.* doi: 10.1080/02723638.2019.1677990.

De'Ath, Amy. 2018. "Reproduction." In *The Bloomsbury Companion to Marx,* edited by Andrew Pendakis, Imre Szeman, and Jeff Diamanti, 395–404. New York, NY: Bloomsbury Academic.

Demos, T. J. 2018. "Blackout: The Necropolitics of Extraction." *Dispatches* 1. dispatchesjournal.org/articles/blackout-the-necropolitics-of-extraction/.

Dillon, Grace L. 2012. "Imagining Indigenous Futurisms." In *Walking the Clouds: An Anthology of Indigenous Science Fiction,* edited by Grace L. Dillon, 1–12. Tucson: University of Arizona Press.

Dillon, Grace L. 2016. "Introduction: Indigenous Futurisms, Bimaashi Biidaas Mose, Flying and Walking towards You." Extrapolation 57, nos. 1–2: 1–6.

Estes, Nick. 2019. *Our History Is the Future : Standing Rock versus the Dakota Access Pipeline, and the Long Tradition of Indigenous Resistance.* London: Verso.

Fairchild, Denise, and Al Weinrub. 2017. "Energy Democracy." In *The Community Resilience Reader: Essential Resources for an Era of Upheaval,* edited by Daniel Lerch, 195–206. Washington, DC: Island Press/Center for Resource Economics.

Gilpin, Emilee. 2017. "An Indigenous Sisterhood Empowers Solar Panel Solutions." *National Observer*, November 3. www.nationalobserver.com/2017/11/03/indigenous-sisterhood-empowers-solar-panel-solutions.

Hall, Rebecca Jane. 2016. "Reproduction and Resistance: An Anti-Colonial Contribution to Social-Reproduction Feminism." *Historical Materialism* 24, no. 2: 87–110.

Jackson, Emily. 2020. "Canadian National Railway Braces for 'Significant' Financial Hit from Rail Blockades." *Financial Post*. March 3. business.financialpost.com/transportation/rail/cn-calls-back-most-of-450-laid-off-employees-to-work-as-rail-blockades-die-down.

Johnson, Bob. 2019. *Mineral Rites: An Archaeology of the Fossil Economy*. Baltimore, MD: Johns Hopkins University Press.

Laboucan-Massimo, Melina. 2017. "Lessons from Wesahkecahk." In *Whose Land is it Anyway? A Manual for Decolonization*, edited by Peter McFarlane and Nicole Schabus, 36–41. Vancouver, BC: Federation of Post-Secondary Educators of BC.

Laboucan-Massimo, Melina. 2020. "Tiny House Warriors." *Sacred Earth Solar*. sacredearth.solar/thw (accessed February 11, 2020).

LeMenager, Stephanie. 2014. *Living Oil: Petroleum Culture in the American Century*. Cambridge, MA: Oxford University Press.

Lennon, Myles. 2017. "Decolonizing Energy: Black Lives Matter and Technoscientific Expertise amid Solar Transitions." *Energy Research and Social Science* 30: 18–27.

Lorenz-Meyer, Dagmar. 2017. "Becoming Responsible with Solar Power? Extending Feminist Imaginings of Community, Participation, and Care." *Australian Feminist Studies* 32, no. 94: 427–44.

MacLeod, Andrew. 2020. "RCMP Accused of Creating 'Crisis of Press Freedom' in Wet'suwet'en Raids." *Tyee*, February 14, 2020. thetyee.ca/News/2020/02/14/RCMP-Accused-Crisis-Press-Freedom/.

Malm, Andreas. 2016. *Fossil Capital: The Rise of Steam Power and the Roots of Global Warming*. New York, NY: Verso.

National Inquiry into Missing and Murdered Indigenous Women and Girls. 2019. *Reclaiming Power and Place: The Final Report of the National Inquiry into Missing and Murdered Indigenous Women and Girls, Volume 1a.*

Pasternak, Shiri, and Tia Dafnos. 2018. "How Does a Settler State Secure the Circuitry of Capital?" *Environment and Planning D: Society and Space* 36, no. 4: 739–57.

Peters, John Durham. 2015. "Infrastructuralism: Media as Traffic between Nature and Culture." In *Traffic: Media as Infrastructures and Cultural Practices*, edited by Marion Näser-Lather and Christoph Neubert, 29–49. Boston, MA: Brill Rodopi.

Power to the People. 2020. "Little Buffalo." *APTN*, January 28, episode 1, 22 minutes. www.aptn.ca/powertothepeople/episode-guide/season-1/.

Riofrancos, Thea. 2019. "What Green Costs." *Logic Magazine*, no. 9 (December). logicmag.io/nature/what-green-costs/.

Runyan, Anne Sisson. 2018. "Disposable Waste, Lands, and Bodies under Canada's Gendered Nuclear Colonialism." *International Feminist Journal of Politics* 20, no. 1: 24–38.

Simpson, Leanne Betasamosake. 2017. *As We Have Always Done : Indigenous Freedom through Radical Resistance*. Minneapolis, MN: University of Minnesota Press.

Spice, Anne. 2018. "Fighting Invasive Infrastructures: Indigenous Relations against Pipelines." *Environment and Society* 9, no. 1: 40–56.

Stefanelli, Robert D., Chad Walker, Derek Kornelsen, Diana Lewis, Debbie H. Martin, Jeff Masuda, Chantelle A. M. Richmond, Emily Root, Hannah Tait Neufeld, and Heather Castleden. 2018. "Renewable Energy and Energy Autonomy: How Indigenous Peoples in Canada Are Shaping an Energy Future." *Environmental Reviews* 27, no. 1: 95–105.

TallBear, Kim. 2019. "Badass Indigenous Women Caretake Relations: #StandingRock, #IdleNoMore, #BlackLivesMatter." In *Standing with Standing Rock: Voices from the #NoDAPL Movement*, edited by Nick Estes and Jaskiran Dhillon, 13–18. Minneapolis, MN: University of Minnesota Press.

Watson, Julia. 2019. "Introduction: A Mythology of Technology." *Lo-TEK. Design by Radical Indigenism*, edited by Julia Watson, 16–23. Cologne: Taschen.

Wenzel, Jennifer. 2020. *The Disposition of Nature: Environmental Crisis and World Literature.* New York, NY: Fordham University Press.

Wilson, Sheena. 2017. "Energy Imaginaries: Feminist and Decolonial Futures." In *Materialism and the Critique of Energy*, edited by Brent Ryan Bellamy and Jeff Diamanti, 377–411. Chicago, IL: M-C-M' Press.

Wilson, Sheena, Adam Carlson, and Imre Szeman. 2017. "Introduction: On Petrocultures: Or, Why We Need to Understand Oil to Understand Everything Else." *Petrocultures: Oil, Politics, Culture*, edited by Sheena Wilson, Adam Carlson, and Imre Szeman, 3–20. Montreal, QC: McGill-Queen's University Press.

Mark Simpson and Imre Szeman

Impasse Time

*E*nergy transition—the shift from dirty to clean energy, from the curse named *oil* to the gift called *solar*[1]—has become a mantra for the present: arguably *the* key phrase by which we name and so perform our response to the impact of fossil fuel use on the planet (see United Nations 2018). The logic contained in the phrase is simple, direct, and (supposedly) easy to grasp. Since the dirty energy of fossil fuels has played (and continues to play) a significant role in generating dangerous levels of carbon dioxide, the use of clean energy will necessarily generate a better outcome for the planet than the use of its dirty counterpart. Yet the broad and deep dependence of existing physical and social systems on fossil fuels, and the need to bring an enormous amount of clean, green energy online globally, together mean that the energy switch must take some time, occurring in measured not frantic fashion—that is, *transitionally*. Even in the face of the enormous challenges posed by global warming to communities around the world—and the need to act as quickly as possible to avoid environmental and social tragedies—energy transition as mantra and logic thus offers a reassuringly temperate, pragmatic, even *Realpolitik* answer to the question of what can be done to mitigate the

The South Atlantic Quarterly 120:1, January 2021
DOI 10.1215/00382876-8795730   © 2021 Duke University Press

unfolding eco-crisis. For these reasons, the phrase names an approach to global warming that a range of actors—environmental groups, broad segments of the public, governments and governmental agencies, international associations, and even oil giants, many of them now mutating into corporate brokers of energy writ large—find palatable, tolerable, comforting . . . and marketable.

The language of transition contains within it all manner of assumptions, some more obvious than others. The fact that we are fated to live with existing levels of atmospheric carbon dioxide for millennia might intimate that any further use of dirty energy, and any further expansion of dirty energy infrastructure (e.g., pipelines, refineries, airports, highways, skyscrapers, suburbs), should stop immediately. Against the apparent unfeasibility of this response, which is treated as either irrational or far too radical, transition proposes a period of phase out and phase in—less the immediacy of "keep-it-in-the-ground" movements than the protraction of urban planning and boardroom decisions. Transitional change is orderly change: a measured, serene, reassuring response to the ragged urgencies of climate crisis. As such, transition must establish clear boundaries and limits. Change the type of energy we use, yes, but make sure to leave other potential alterations and alternatives unthinkable: managed, diminished, foreclosed, or stifled altogether. For the most part, the logic of energy transition thus presumes that *absolutely everything else*—and especially neoliberal capitalism, its structures, practices, and protocols, and the vast inequalities of power and privilege they generate—will stay much the same. If *solar* is renewable energy's synecdoche, then *the solar fix* gives one name for the transitional sameness we are describing. We posit that solar-as-fix contradicts and betrays whatever *solarity* might turn out to mean.

As an idiom and a grammar, transition is intended to address the looming threat that environmental crises pose to current modes of power by draining away the energies of opposition over time, while in the process updating neoliberal governance with the shiny surfaces of renewable energy. Key here is transition's autopoetic aura, through which it promises quite magically to realize itself and thereby guarantee the happiest of happy outcomes. And such magic serves while also charming the human: the language of transition insists on humanity's preeminence and views ecological calamity as a problem to be addressed so that humans can continue to live on the planet. Global warming has raised fundamental challenges to ontology and epistemology, to ethics and aesthetics, to understandings of the non-human, and more. By contrast, exponents of transition-logic have little time for the deep questions that come to light when gazing over the cliff of climate change, preferring to believe that all we need is to get on with the business of building a LEED Platinum-certified bridge across the abyss.

Energy transition thus constitutes a meek response to global warming, a scheme that status quo economic and political actors appear glad to take up as a way to save themselves from the worst of what is to come: ostensibly the ardors of climate catastrophe, but really the seismic shocks of stock market collapse, currency devaluation, resource redistribution, and the like. After all, an electric car to replace the gas-guzzler still leaves a private vehicle in every driveway and, for car manufacturers ready to make the transition, a profit in the bank. Yet the very currency of energy transition as a mantra, not to mention the relatively timid and innocuous energy changes it demands, mean that *energy impasse* is actually the defining condition of our time. Even as scientific consensus overwhelms us with evidence about the environmental consequences of oil societies, petroculture still persists and—with every new ring road, pipeline, and fracking rig—redoubles. The use of dirty energy is a reality and a concrete problem that demands our collective attention. So too does the sedimented, intensifying condition of energy impasse—a more complicated, more abstract, and perhaps ultimately more dangerous figuration of the fossil fuel era alongside its transitional supplement that requires reckoning in units of measure other than parts per million of carbon dioxide.

What do we mean here by *impasse?* For of course there are all manner of obdurate problems impeding meaningful changes in energy use, problems that slouch and brood, impassively, on the contemporary landscape. Consider, for example, the rightward tilt signaled by Trumpism and Bolsonarism: manifestly a form of impasse. The recent turn of a number of nation states (or of provinces and states within them) away from environmental policies designed to provide ecological protection and to limit the generation of pollutants and greenhouse gases underscores the hollowness of transition rhetoric. For every small country that has made commitments to wean itself off fossil fuels, there is a large one like Brazil, whose policies seem designed to deliberately upend the atmospheric apple cart, or Canada, whose lack of policies stands to have much the same impact (Cunha 2019; Rabson 2019). Or consider the privileged givenness of fossil-fueled mobility for so many in the world today, as indexed by the unrelenting expansion of automotive culture: global car sales have hovered between seventy-five and eighty million over the last five years—a 45-percent increase over average sales of fifty-five million units between 2000 and 2015—and there have been over *one billion* cars sold this century (Statista). Or consider, more abstractly, the double bind articulating the environmental urgency of energy transition (it must happen now!) against the social difficulty of that transition (how might it happen at all?). We could add many more problems to this list. Every one of them constitutes a clear impasse that prevents changes in our energy system—and therefore a pressing concern for any viable politics today.

All the same, we invoke impasse here with something a bit different in mind. We see impasse not so much as *blockage,* that is, an impediment to a given situation that requires circumventing or dissolving or overcoming. Instead, we understand impasse as *stuckness*: the texture or atmosphere setting the conditions of possibility for a given situation that, irrespective of any overcoming of actually existing blockages, manages nevertheless to perpetuate the situation as it is. Impasse in this sense names a continuation of the same wherein the overcoming of blockages cannot solve—and may in fact compound—the abiding stuckness. We are reminded of Fredric Jameson's (1982: 153) memorable claim that "the deepest vocation" of the utopian genre "is to bring home our constitutional inability to imagine Utopia itself." Genres of energy transition in the current moment operate in much the same way: less to provide viable means for a better future than to indicate our constitutional inability to imagine transformation itself and thus manifesting the conditions of our stuckness. Which is to say that existing genres of energy transition are all too often forms of impasse.

The stuckness to which we want to draw attention is conditioned by a narrative at the heart of transition's logic: that there have been *other* energy transitions, that the time of energy is *always* about transition. The dirty energy era began, the story goes, with an originary shift from wood to coal and, after successive periods of realignment in the dominant forms of energy used, will find its apotheosis in the transition from oil to clean energy. The conjunctive sequence from wood to coal to oil to nuclear animates this broadly progressive narrative, which is partially why it is presumed that what has to come next are renewables (though how sun and wind energy are more advanced than nuclear as sources of energic power troubles the narrative). The prospect of transition is also imagined as if coordinated by a single sovereign entity: a global principle—whether "technological progress" or "the market" or both—that benignly and beneficently can coordinate everything.

Like so many other narratives governing our practices, this one, too, turns out to be a fiction. In *The Shock of the Anthropocene,* Christophe Bonneuil and Jean-Baptiste Fressoz (2017: 101) write:

The bad news is that, if history teaches us one thing, it is that there never has been an energy transition. There was not a movement from wood to coal, then from coal to oil, then from oil to nuclear. The history of energy is not one of transitions, but rather of successive *additions* of new sources of primary energy. . . . Energy history must therefore free itself first of all from the concept of transition. This was promoted in the space of politics, media and science precisely so as to spirit away worries bound up with the "energy crisis," an expression that was then [in the early '70s] still dominant.

What emerges from Bonneuil and Fressoz's account of the messy reality of addition against and instead of the antiseptic fiction of transition is a crucial question of precisely how transition is supposed to take place at all. If history provides no example of transition, then any guiding principle of transition is merely speculative or notional at best. So why does its fiction persist? Precisely to normalize and naturalize the logic of transition as given, as just over the horizon, despite all impediments (for how could Trumpists and Bolson-aroites stand in the way of inevitable progress-to-come?). Transition's fiction thereby makes sure that we remain stuck in a present that withholds, in the active language of its idiom, any capacity to create a genuinely different energy future.

To the extent that *solar* designates or indeed epitomizes the transitional given today—the reflex postulate or preordained synecdoche condensing within its name all hopes for energy futurity—it arguably consolidates instead of disturbing the condition of impasse. As the code word for all manner of new energy forms, practices, and relations on the horizon, solar makes us give over those unnamable, incoherent desires, impulses, and demands that might otherwise confound or discombobulate transition's impassive logic. Enthralled by solar's synecdoche, we forget to remember the questions we might want to ask about the narrative of energy transition and, in so doing, we settle for the given by reaffirming the progressive script encouraging us to maintain and to trust that it was always going to be solar all along. This peculiar circumstance constitutes *the solar fix*: a binding condition not just distinct from but antithetical to the promise of solarity.

Time and Impasse

Before we venture to unpack this claim, we want to dwell a bit longer with impasse in order to reckon, more particularly, the significance of time for impasse as a problem. The concepts of *present* and *future* we have been tracking raise in turn the issue of temporality for the matter of impasse today. If, as we assert, energy impasse is the defining condition of our time, then what is the time of impasse? What temporalities does impasse seed, and how might a reckoning with those temporalities afford us some position or perspective against the pull of stuckness? We venture that impasse has everything to do with time in relation to how we think futures, and so that the logic of transition (in energy systems as well as in ecological and environmental sensibility) has everything to do with futurity and the modes of its imagining.

Let us explain in more detail what we mean with respect to the time of impasse. We are inspired in our thinking by the compelling argument advanced by Timothy Mitchell in his 2014 article "Economentality." Tracking the emergence of "the economy" as a discrete object in the global imaginary in the years immediately after the Second World War, Mitchell drills down in particular on the decisive significance of this new object for orders of time within liberal governance. "The economy," he contends,

> provided a more pervasive effect, one that has since then escaped attention: a way to bring the future into government. The appearance of the economy established a new temporal scheme in which past, present, and future were relocated. We can follow this shift . . . as a new prognostic structure in which a future was mobilized as a mode of adjudicating and managing claims in the present. The government of the present, as it was imagined through new forms of the future, would come to operate within a new metric of temporal change, the measurement of growth. (484)

Here the newly figured autonomy of the economy, as a discrete object that would grow, establishes a novel time signature for the work of governance: economic expansion into the future becomes an aim in and of itself.

For Mitchell, the economy as socio-political effect—what he calls economentality—served to address two conjoined problems: labor struggles disrupting global energy relays and the specter of limits that had haunted the interwar period, when blockage to growth increasingly seemed endemic to capitalism. Against labor's agitations from below and against the Keynesian common-sense that capital had reached its limit, the economy as autonomous effect offered sovereignty a new opening for governance: a means of "embed[ding] people's political lives in the future by bringing them to calculate according to its representation" (Mitchell 2014: 492). Thus, economentality fashioned limitlessness as the retooled fantasy of liberal progress, positing unfettered material and temporal plenitude as its future horizon. In this fantasy, eventually everyone will have everything—a promissory logic that operates through deferral, whereby most subjects subsist in perpetual anticipation of the eventuality-to-come that never in fact arrives.

The lifespan for this mode of governance in its proper, functional incarnation was, by Mitchell's accounting, actually quite brief: "The economy worked effectively as a mode of government-through-the-future for only a couple of decades. By the late 1960s, the forms of productivity growth, energy use, cheap oil, and Middle Eastern politics on which it depended were all under pressure" (507). The ensuing shocks—often figured via the shorthand 1973—could, especially in view of the concomitant rise in environmental awareness and ecological commitment, have forced a shift in direction guided by the recognition that the mid-century discourse on the economy and economic futurity was not working. What emerges instead, however, is a desperate drive to hold on, one manifest not just in geopolitical retooling of the ways in which the US and other global northern powers access energy but also in the birth of the so-called New Economy, through which technology comes increasingly to supply the magical solution to economic and energic crises simultaneously.

The point we would emphasize, in composing this quick genealogy, is that what we call impasse, so apparently specific to the contemporary moment, actually has a quite prolonged emergence. It proves contiguous with—and constitutes one dimension of—neoliberalism as an order and ideology of governance. Thus while the present conjuncture, overloaded with dire signs of limits breached and futures ruined, must seem very different from the mid-century moment when progressive plenitude reigned serenely supreme, we would nonetheless argue that impasse today enables the "temporal scheme" of economentality to persist in fractured form. Impasse retrofits plenitude to keep neoliberal governance going. Even as there is now an alertness to the limits of the narrative of perpetual progress, the stuckness of impasse perpetuates this narrative all the same, precisely by holding in suspension the promise of some sustainability beyond the impasse while pitching techno-utopian solutions to the problem that will take time—that must unfold across an eventual horizon.

Sustainability's Suspense

Put another way, sustainability serves the neoliberal retrofit of plenitude by synching transition-logic to austerity's more-with-less mandate. As a concept and a grammar, sustainability, so commonly used in discourse on energy and environment, thus bears in significant ways on the problems of time and impasse we have been examining so far. If *the economy* was the mode of governing for the future for the post-war era, *sustainability* is the mode for the present moment, one in which other limits need to be accounted for alongside those of wealth and value. And what we have termed the solar fix is the energic form of sustainability as a mode of governing in this sense.

In "'After the Sublime,' after the Apocalypse: Two Version of Sustainability in Light of Climate Change," Allan Stoekl (2013) probes the limits of the idea of sustainability as a concept adequate to initiate energy shift, transition, or revolution. He offers a critique of sustainability that has by now become somewhat standard within eco-criticism. What exactly is the metric of sustainability? For precisely whom is the planet to be made sustainable (only for humans and not other species)? What exactly is the timescale of sustainability? Fossil fuels are of necessity unsustainable, which demands the creation of renewable forms of energy and energy infrastructure. But is it possible to conceptualize how much renewable energy might be sustainable? Responding to the imperative to act more sustainably put forth by the 1987 Brundtland Commission Report, "Our Common Future," Stoekl (2013: 48) writes:

> Brundtland, despite its seeming certainty, inevitably gives rise to multiple possibilities and scenarios; the future of sustainability begins to seem less like a clear roadmap of choices than a menu of possibilities, a panoply of fictions that operate on both the aesthetic and the moral plane as well as on the "practical." In that sense sustainability is both a life-and-death matter and a literary—and literary-theoretical—practice.

It might be tempting to thus drop sustainability as a discourse that navigates the future in a manner more attuned to environmental limits. The intent of a document like "Our Common Future" is, first and foremost, to safeguard the environment only by ensuring that the open horizon of capitalist growth and liberal notions of progress remain in place via sustainability as regulatory ideal (sustainability is never linked to ideas of degrowth). However, Stoekl decides to try to re-narrate sustainability rather than abandon it, in large part because it is a concept that draws together the issues that animate environmental politics: the status of the human, the fate of the environment and resources, new understandings of community, and time. He proffers two kinds of sustainability: a first-order, general sustainability and a second-order, restrained mode of sustainability.

General sustainability constitutes a limit case. The human is just one species among others, a species that (given its practices and actions) might persist or not. The world is sustainable no matter what humans do to it, even if it might not be sustainable for humans themselves. There is coded into the language of sustainability a humanism and an imperative for human survival that general sustainability decisively calls into question. While this might be interesting philosophy, we cannot help but characterize it as bad politics—a ceding of the earth to extant practices of capitalism, an extreme accelerationism that results not in radical social change but in the disappearance of the social altogether.

As a potential response to the stuckness of impasse, Stoekl's account of restrained sustainability is more intriguing. As elaborated by Kant, the sublime constitutes something of an epistemic parlor trick: scale unnerves sensibility only to reaffirm in the end the absolute ability of human cognition to understand even that which appeared to be impossible to grasp. The sublime that Stoekl references in the title of his paper works differently than this: when it comes to trying to calculate the vast array of externalities that would make up a sustainable society, our cognition cannot help but falter. There is no overcoming of the sublimity induced by sustainability. Instead of leading to a discounting or disinterest in life in the future and a dangerous affirmation of present life (which amounts to either a cynicism about the environment or an apocalypticism), Stoekl sees a potential opening for a different form of sustainability. This is a mode of sustainability organized around a demand for the future: "A certain world, a certain climate, a certain human population, a certain ecology with certain animals" (Stoekl 2013: 48). In restrained sustainability, narrative, morality, and the affirmation of community constitute devices that simultaneously affirm the need for something like sustainability, but recognize that it is ultimately unrepresentable. In brief: the dangerous self-certainties of sustainability in "Our Common Future," which help to foster a hope that we can continue along with a slightly muted, slightly slower version of capitalist growth, are suspended; in their place, the moral and political imperatives advanced by the sublimity of sustainability generate life practices "stripped of all illusions concerning that very sublimity" (50). Restrained sustainability doesn't anticipate the future, but takes up the ongoing, endless, and contested challenge of writing and enacting it.

We don't think either general or restrained sustainability, provocative though they are, constitutes an adequate response to the challenge posed by impasse. But note how time works in each of these versions of sustainability. The Brundtland Report names environmental limits in order to render them unimportant: the right policy decisions paired with technological advances allow the extraction of value to continue unabated. For Stoekl, the illogic of sustainability (i.e., what could the metric possibly be for a complex and changing ecology?) generates two responses. General sustainability poses an end to the human instead of a continuation—an eschatology in place of history. Restrained sustainability opts for neither ends nor temporal continuity, but operates in a suspended time, organized around "tactics linking aesthetics, technics, (base) materialism, and fiction, which are embraced not as absolute meanings but as memes, finite structures of meaning, connected to survival practices and tactics" (54). What is so cunning about the ruling mode of energy transition is its capacity to render such suspended time itself lucrative for capital. It promises a future in which we might engage in the challenging sociality of restrained sustainability, while in fact pushing it far off into a distant future that might never come: given enough time, technology will domesticate the sublimity of sustainability, rendering null and void all the questions it cannot help but raise.

Solar offers perhaps the *ur*-example of this misplaced faith in technology, although it operates through a slightly different relation to time. There is a reason why solar has all too often been viewed as a solution not just to the use of dirty energy, but to *all* of the troubles of the social. In an unrecognized confirmation of the insights of the energy humanities about the deep links between energy and social form, solar is imagined as fully unsettling the apparent rationality and presumptions of petroculture through the temporal update it performs. Solar technology produces energy from sunlight that is minutes old, rather than relying on the ancient sunlight collected in fossil fuels: it provides the ultimate update to history, making energy fully present to the present. But there is another trick of time that solar performs. Solar energy is limitless and timeless, characteristics that allow it to negate the threat of general sustainability while rendering the difficult trade-offs of restrained sustainability unnecessary, beside the point: there is more than enough energy to go around for everyone. By negating both of the modes of sustainability outlined by Stoekl, solar puts the human back at the center of history and allows it once again to be all too human: back in the game of doing pretty much what it wants. This indefinite, timeless, limitless realm, turning on a form of suspended time that proves particularly difficult to parse, sounds more like a description of the deepest fantasies of capitalism than some new mode of sociality that might attend to the non-human and to the innumerable limits that exist outside of and beyond energy (soil and food, water and minerals, and all the rest). Put more bluntly: as time signature, solar timelessness converts the solar promise into the solar fix.

Beyond Impasse Time; or, the Impossible

We hope we have made compelling how significant time is in figuring, for political purposes, the need to act on climate crisis. One of the most powerful arguments for us on this score is Andreas Malm's description, in *The Progress of This Storm* (2018), of what we would call disjunctive belatedness: climate change as the revenge of history on the present; long-burnt carbon arriving now to imperil any version of the future.[2] This devastating account of the convoluted temporality in which we dwell obliterates any trace of plausibility that progressivist narratives might claim to offer. And yet somehow such narratives persist in enthralling us.

How might we interrupt the hold on time that energy impasse exerts, a hold inextricable from the one we find in these enthralling progressivist narratives? It is the temporality of impasse that generates a belief in the untrue. Undoing such belief means interrupting that temporality. Will the urgency of the immediate present, the imperative to confront the prospect of *now*, provoke us to achieve such interruption? Impasse leaves us stuck in a broken, unworkable present—a cancelled now—through its many fantasias on futurity. The order of the now contradicts impasse by wrenching us from present paralysis. As we have been arguing, the temporal schema of impasse is the cynical retooling of the progressive, eventual one. It seems to us that, by confronting the urgencies of *now time*, we might supplant this progressive temporal schema with a new time signature: not past-present-future but, instead, the now and the next.

At the moment we are writing, the now and the next seem bleak: they involve terrors of plague and trials of quarantine. Yet even in the teeth of such disorienting shocks, the progressive temporality of energy transition proves obdurate, impassive. On 19 March 2020, amidst the global uncertainty caused by the COVID-19 pandemic, BP proudly announced: "Lightsource BP completes financing on 260 MW solar project in Texas." According to an executive VP quoted in the press release, "This project demonstrates that the competitiveness of solar energy means that power offtake structures widely and historically used for conventional generation are now gaining traction for solar energy projects. We see an exciting future from the increase in competitive renewable energy in the US power markets" (BP 2020).[3] The retooling of fossil infrastructure ("power offtake structures widely and historically used for conventional generation," in BP's catchy euphemism) for a dawning solar era that will deliver the exciting futurity of ever more competitive power markets epitomizes exactly the tendency we observed near the outset: that as *solar* names the transitional given today it merely consolidates instead of disturbing the condition of impasse. And while presumably Light-source BP's excellent Texas ad-venture has been in the works since long before novel coronavirus arrived to disrupt everything, it remains difficult to resist reading the announcement and its timing as a pointedly *sunny* over-correction for the perils of the now and the next by way of the solar fix.

What might distinguish *solarity* from the solar fix? The matter, we imagine, is one of orientation or, indeed, of what Keller Easterling (2014: 21) calls "disposition." In her keynote at After Oil School 2, Nicole Starosielski reminded those in attendance that "we do not look at the sun itself"—not least because to stare sunward is literally blinding (2019).[4] This disarmingly simple insight implies that the solar fix, fixating (us) on the sun as key to the perpetuation of futurity-as-progress, is a species of blindness. Solarity, by contrast, turns on a glaring and productive contradiction: that troubling to look away from the sun so as to concentrate on social solidarities might actually allow some surprising solar alternatives against and beyond the solar fix—and with them some unforeseen social relations impossible within the stuckness of impasse—to begin to come into view.

Notes

1   In our usage *solar* serves as a shorthand for renewable energies of all kinds, including wind, tidal, geothermal, and hydro, even though we also want to insist on the specificity of the solar as the key sign and symbol of future energy.

2   "There is no synchronicity in climate change. Now more than ever, we inhabit the diachronic, the discordant, the inchoate. . . . History has sprung alive, through a nature that has done likewise. . . . Postmodernity seems to be visited by its antithesis: a condition of time and nature conquering ever more space. Call it *the warming condition*" (Malm 2018: 11).

3   Do not suppose that BP is oblivious to the challenges of the present moment—its executives are nothing if not grounded realists, as the press release's concluding section, "Working safe and smart," makes clear: "At Lightsource BP, the health and well-being of our team members and partners is our top priority. We are actively monitoring updates regarding the novel coronavirus (COVID-19) and are following precautions and guidelines provided by the CDC and public officials" (BP). Under plague conditions, does working safe and smart to advance green growth and ensure transitional sameness exemplify restrained or general sustainability?

4   The claim of which this reminder was a part, while less literal than our riff here, drives home the larger point about disposition we are making: "Solarity can be most transformative when we do not look at the sun itself or the sun as an energy source" (Starosielski 2019).

References

Bonneuil, Christophe, and Jean-Baptiste Fressoz. 2017. *The Shock of the Anthropocene: The Earth, History and Us.* Trans. David Fernbach. London: Verso.

BP. 2020. "Lightsource BP completes financing on 260 MW solar project in Texas." *bp.com,* March 19. bp.com/en/global/corporate/news-and-insights/press-releases/lightsource -bp-completes-financing-on-260-mw-solar-project-in-texas.html.

Cunha, Daniel. 2019. "Bolsonarism and 'Frontier Capitalism.'" *The Brooklyn Rail.* February 5. brooklynrail.org/2019/02/field-notes/Bolsonarism-and-Frontier-Capitalism.

Easterling, Keller. 2014. *Extrastatecraft: The Power of Infrastructure Space.* London: Verso.

Jameson, Fredric. 1982. "Progress Versus Utopia; Or, Can We Imagine the Future?" *Science Fiction Studies* 9, no. 2: 147–58.

Malm, Andreas. 2018. *The Progress of this Storm: Nature and Society in a Warming World.* London: Verso.

Mitchell, Timothy. 2014. "Economentality: How the Future Entered Government." *Critical Inquiry* 40, no. 4: 479–507.

Rabson, Mia. 2019. "Canada's Emissions Target Gets Further Away As 2017 Report Shows Increase." *Globe and Mail,* April 16. www.theglobeandmail.com/canada/article-cana das-emissions-target-gets-further-away-as-2017–report-shows-2/.

Statista. 2020. "Number of Cars Sold Worldwide from 1990 to 2020." *Statista.com.* statista.com /statistics/200002/international-car-sales-since-1990/ (accessed August 12, 2020).

Starosielski, Nicole. 2019. "Harvesting Sunlight." Keynote, After Oil School 2: Solarity, Canadian Centre for Architecture, Montreal, Quebec, May 24.

Stoekl, Allan. 2013. "'After the Sublime,' After the Apocalypse: Two Versions of Sustainability in Light of Climate Change." *Diacritics* 41, no. 3: 40–57.

United Nations. 2018. "Sun's Energy at Centre of Revolution in Renewables, Secretary-General Tells International Solar Alliance's First General Assembly." UN SG/SM/19272– ENV/DEV/1891. October 2. un.org/press/en/2018/sgsm19272.doc.htm.

Amanda Boetzkes

Cold Sun • Hot Planet:
Solarity's Aesthetic, Planetary Perspective

William Blake asked the tiger: 'In what distant deeps
or skies burned the fire of thine eyes?' What struck
him this way was the cruel pressure, at the limits of
possibility, the tiger's immense power of consumption
of life. In the general effervescence of life, the tiger
is a point of extreme incandescence. And this
incandescence did in fact burn first in the remote
depths of the sky, in the sun's consumption . . .
—Georges Bataille, *The Accursed Share: An Essay on
General Economy. Volume 1 Consumption* (1991)

This quote from Georges Bataille summarizes a
nexus of thinking about the general economy of
the sun; how its energies compel planetary life
with a need to consume, accumulate, and expend
surplus energy in ways that ensure its movement from one being to another and through
consciousness itself. But, for Bataille, this energy,
and our consciousness of it, is rife with ambivalent reversals that trouble ontological distinctions.
For the sun's energies chain human beings to
our mortality, animality, and importantly for
this article, *planetarity*. The sun's reversals—its
nourishment of beings and its violent expenditure
of them in what Bataille calls its "depredations of
depredators"—cannot be understood through the
physics of energy exchange, nor even its politics

The South Atlantic Quarterly 120:1, January 2021
DOI 10.1215/00382876-8795742   © 2021 Duke University Press

per se (Bataille 1991: 34). Rather, its ambivalent movement perturbs these systems of knowledge altogether.

Bataille's account of solarity from his early essays in the nineteen-twenties and thirties to his three-volume speculative theory of economy in *The Accursed Share* (1946–49) ventures through figurations of the pain of accumulation, the sadism of expenditure, and the catastrophic expulsions of the earth as these are all envisioned from the sun's perspective in the depths of the sky. In short, solarity is a lens by which to see the movement of the sun through planetary beings in processes of growth, consumption, death, and decay. In the epigraph with which I open, solarity is mirrored as a hunger that burns in the tiger's eyes and a cruel pressure that invests the tiger with the energy to consume. In this vein, solarity is a potential energy that passes through planetary life: it compels the tiger to hunt its prey, but it will also push through the tiger's body, driving it towards its own death, to rot into the earth and return to the raging heat of competition for the sun's nourishment once again.

The tiger is but one of many allegories by which Bataille articulates human solarity as a fundamental desire to consume and its capacity for destructive energy expenditure. Solarity, it seems, is the same no matter what being it occupies: it drives all beings to expand to the point of their own death, a passage and absorption into other beings or material states. In this sense, solarity is a fantasy of the death of capitalism and an end to its myth of self-expansion. Solar energy produces a self-expending drive that condemns the bourgeois capitalist economy to disburse itself into base matter. Most importantly, Bataille pursues this fantasy by situating capitalist subjects and relations in painful struggle with their planetarity, the elemental forces of cold and heat, ice and fire, blinding light and nocturnal depths that disfigure the human body.

In what follows, I address Bataille's conjoining of solarity and planetarity with an attentiveness to the plenitude of his writing and its painful effects. Thinking solarity's potentials alongside mass extinction—the end of lives and the possible end of life itself—is dreadful. It calls forth violently contradictory reactions at the limit and possibility of the human, as well as the limit and possibility of the social, the collective, the Other more broadly, and others in particular. While it is tempting to think solarity as the energic condition for a redistribution of being, subjectivity, and collectivity, this cannot take place without a struggle with the frightening potential of contemporary capitalism and its resourcing of life as energy *as the very form solarity might take.* I recuperate Bataille's solarity, then, in order to articulate the risk that solarity might mutate with the petrocultural regimes from which it emerges and become a new capitalist energy regime that accelerates or otherwise exacerbates our current course toward planetary destruction. The question I would invite us to consider in this reconsideration of Bataille's planetary aesthetics is, can we expend the sadism implicit in capitalism's anticipation of the future to take hold of solarity otherwise?

Solar Phantasms: Writing Capitalism at the Boiling Point

In *The Accursed Share*, Bataille considers the willful destruction of property in the Northwest Coastal First Nations' practice of the potlatch and of Aztec human sacrifice to the sun as complex forms of energy expenditure and key instances of solarity in operation (1991). In his analysis of this text, Jean Baudrillard makes the important point that Bataille figures Aztec solarity as a generous economy that derives from the knowledge that the sun *gives nothing*: "The unilateral gift does not exist . . . it is necessary to nourish it continually with human blood in order that it shine" (Baudrillard 1998: 193). Solarity is not a pure form of generosity, then, but rather the economic extension of the sun's insatiable demand for life. This is why Bataille teases out a bi-directional relation from the concept of consumption: it is at once the excess energy that must be expended and the excessive desire to consume in order to generate that excess energy. The incandescence of the sun gives generously but equally burns within living organisms as a demand to be fulfilled, like the predatory hunger in the tiger's eyes.

In this way, Bataille's account of solarity mirrors the logic of capitalism with its implicit demand for sacrifice. His apparent celebration of "solar societies" must therefore be carefully read in terms of how his language enacts the double meaning of consumption: it both posits the restrictions of capitalism and drives against them. Thus, for Baudrillard, the sun's expenditure of energy is not implicitly generous; it is not the site of solarity. Rather solarity shines through Bataille's writing as it positions a *"subject of knowledge always at the boiling point"* (194, emphasis mine). His dazzling visions evoke mythic forces and figures that he sets against disciplinary formulations of objective knowledge, such as Marxism, anthropology, political economy, and scientific method. It is not the sun, but Bataille's mythologization of it in the face of restricted economies of knowledge that is the fulcrum of expenditure.

Bataille recuperates the generosity of solarity through his mythic assertions and their capacity to draw out unthought subject positions. His invented archaic language is directed to the outpouring of dynamic elements into words that disfigure the subject and knowledge itself, through their narrative and graphic unfolding as violent interception and continuous overflow. Bataille's deployment of solarity as a mythic agent is therefore better understood as an operation of destructuration, not just of words or concepts, but of entire systems of knowledge. This operation is what philosopher Rodolphe Gasché calls Bataille's phantasmology: his writing and positioning of myth in such a way as to set it against the repressive scaffolding of science and philosophy (2012). Phantasmology is an anti-science: it raises mythic phantasms such as vital elements, sense-effects, archaic forces, and unintelligible narrative excesses that have been suppressed and expelled in order to achieve the transparency of scientific logic. Bataille's writing sets phantasms against the logic of science and philosophy to "shatter their peace" and outstrip their intelligibility (111).

By unleashing phantasms as ontological forces in his writing, Bataille anticipates the planetary thinking of political ecology. For example, his phantasmology informs Gayatri Spivak's formulation of planetarity as the Urform of alterity; a primary figuration of the disfiguring force of the other (Spivak 2003: 71). Planetarity, she suggests, makes an uncanny appearance that is disjoined from identifications with globalization. To think ourselves as planetary, rather than as global or worldly is to radicalize alterity itself (one's own and others') in ways that are not derived from a global imaginary still entrenched in a colonizing framework. Despite the command of globalization to schematize the imagination, Spivak argues, humans tend toward an alterity, a transcendental figure whether nature, mother, or god, that is attributed with an original animating force. Yet this original force always risks being reabsorbed by the logic of globalization. Planetarity is therefore a continuously receding domain, never bound to the strictures of the figure. It is both a radical figure of alterity in contrast to "nature", but also an insistent operation of disfiguration by which the planet and planet-thought preserves its zone of irreducibility within the dominance of globalization. Planetarity acts at once as a force, an embodied position, and a philosophical disposition toward alterity.

In a similar vein, Bruno Latour, Isabel Stengers and Donna Haraway all introduce a panoply of mythological, classical, or fictional figures into their philosophies of planetary modes of existence as remedy to the epistemes of the modern and to channel the uncanny quality that inflects the intrusive appearance of a planetary reality (Latour 2004; Stengers 2015; Haraway 2016). Each invokes the name Gaia to nominate a mythic being that governs these planetary intrusions. This approach is not to designate a truth (with scientific authority) but is rather a theoretical praxis: a mode and language that is sensitized to planetary forces, and which confers upon us humans the power to experience and think the planet in consonance with its indifference to us. Solarity puts us at the boiling point of planetary thinking. As it threatens to overflow our economy and its correlated knowledge systems, solarity calls us to remember the planet's indifference to us, to that global economy and its forms of knowledge. Its indifference is precisely its generosity.

A Paradoxical Elemental Intimacy with the *Jesuve*

By reading Bataille's solarity next to contemporary theorizations of planetarity (the mythic earth as primary alterity), it becomes possible to discover in solarity the disorganizing force that perturbs the limits of economy and science, and unseat them from their authority over the ebullience of life. But this cannot occur without a radical disfiguration of the subject itself, one that effects a becoming-planetary as its energies deflagrate the subject of knowledge. Here is where solarity is channeled through the aesthetic effects of the phantasm. Solarity becomes an encompassing term for the many earthly and cosmic elements that prey on Bataille's subjects, pressuring them and then pulling them apart in the paradoxical movement between the accumulation of energy and its total expenditure unto infinity.

It is remarkable that Bataille embraces the sun by figuring it as an agent of bodily torment. (This should alert us that solarity is a deeply ambivalent subject matter, meant to be handled with care and not abstracted and theorized as energy culture, or worse, appropriated towards a total and destructive capitalist command of subjectivity). Bataille's early formulation of solarity in the late nineteen-twenties, "The Solar Anus," is a painful exercise in writing his own phallocratic body into desire and death, across the loneliness of ontological solitude (1985: 5–9). The essay moves between his own intimacy with a lover (an eighteen year old white bourgeois girl), and the violent elemental dynamics between the sun and the earth. He and his lover, sun and earth, are bound together in a paradoxical elemental intimacy. They become interpenetrated but also propelled apart into endless isolation and eventually complete annihilation through a form of planetary coitus that cycles across life and death in perpetuity. In other words, the resituating of sexual communion through solarity destines his body and his lover's into a staging of the cosmic rancor between the shadowy depths and energies of the earth (which he figures as specifically anal) and the unyielding exposure of the sun (which he figures as phallic).

Bataille rewrites the dynamics of heteronormative penetrative sex alongside earthly dynamics, like the movement of the tides, sea lapping against the sand, the rotations of the earth and the moon. He shows how these terrestrial rhythms begin to accumulate the energies of planetary elements and overflow under the influence of the sun; how the bodies begin to generate excesses that drive the lovers into deeper depths of the night, specifically, a deathlike sleep from which he awakens as a corpse-like figure. His reawakening is accompanied by an increasingly violent need to violate bodily and topographic boundaries. From the night, and the lovers' sleep of ontological solitude, the sun awakens him to a different, unthinkable intimacy: a drive to sadistically and with accelerating repetition penetrate his lover's anus to the point of mutual annihilation (7–8). The sun is the dynamic force that drives his body into desire and expenditure—as blood-filled erection and ejaculation—and activates a repetition compulsion. The sun does not, will not, and cannot stop: it drives his body's re-emergence as a corpse that continues to desire and expend. The lovers' bodies have intermingled and been transposed into a solarized planet, figured as the fluidity of the earth's rot, soil, and lava spewed from volcanoes. The earth's cracked topography thus becomes the sun's anus: the site of its violent ejection of contents. The sun's generosity is therefore not pure abstract light or heat, but is rather countermanded by the earth's profane and hidden decay. Bataille refuses to think phallic solarity without its carnal opposite, the anal night, which serves to produce a totality of cancelation: the "solar annulus" by which his lover's anus, the earth's anality, and the drive to annihilation are bound up in his writing of the sun.

In this early essay, Bataille initially figures solarity as phallic, sadistic, and predatory. But he cannot write solarity without its opposite: the earth, the (female) anus, the fluidity (blood) with which the earth is replete and which it expels as lava from its volcanic topography. In "The Solar Anus," the sun's energy essentially de-phallicizes the male lover as it "analizes" the female lover. Both bodies are violently stripped of their procreative coding and are recast in the sun's violent exchange with the earth. Ultimately, Bataille gives a name to this excessively phallic but paradoxically de-phallicized (annulled) male figure, which he elaborates in subsequent essays: the Jesuve. The neologism combines the words "Je suis," "Jésus," and "Vésuve" (Vesuvius), to articulate the terms by which solarity intervenes on male subjectivity and its sacred underpinnings in Christian theology. Bataille pollutes the image of the singular, bleeding Jesus on the cross (God as man), who is sacrificed and whose blood is expended to cleanse the world's sin. The

Jesuve is the obscene and godless synthesis of the sun and the earth, a profane Jesus, at once phallic and anal. It is a subject replete with blood that it seeks to expend violently, whose face is hideously red, obscene, and volcanic, and whose words are excrement. Bataille writes, "I am not afraid to affirm that my face is a scandal and that my passions are expressed only by the JESUVE . . . the filthy parody of the torrid and blinding sun" (8–9).

As a parodic image of the sun, the Jesuve derives its power from its own operations of self-expenditure/self-cancellation/self-sacrifice. Solarity is its disfiguring force, a dynamic energy that mutates the human body and its meaning. The sun offers the capacity for heterological readings of the intertwining of the human body, consciousness, and earthly form. In so doing, it becomes possible to position the Jesuve as a transitive energy by which to write solarity and revolutionary energy together as co-extant planetary forces that well up in the human body and living things more broadly conceived. Thus, Bataille recuperates the sun as a dazzling force the rises up *from below,* blazing through the body and bursting through the top of the skull. The sun possesses the body and absorbs it into an anal/annulling, blind/blinding pineal eye. Now, blindness and vision appear together as the sun's maddening and cancelling effect. Blind sight becomes the operative force and condition of visuality itself at the origin of planetary consciousness.

Cold Sun, Blind Sight

Bataille's writing discovers in solarity a mythological vision that stands as the antithesis of Enlightenment philosophy. In this regard, Bataille casts science and philosophy in the same cold light of the sun in a method he calls heterology, or "the science of the heterogeneous," (1985: 97). He imagines a solar light that would counter the presumed objectivity of the Enlightenment tradition. In order to do so, Bataille drew from Friedrich Schelling's concept of the katabole in his *Philosophy of Mythology* (2012). As Rodolphe Gasché explains, the katabole is a tripartite concept that means: (1) to cast down, to throw down, to push down; (2) to originate, to ground, to begin; (3) to throw away, to cast away from oneself (60). It is at once an action, an actant that is cast out from the subject, and a founding concept. Taken together, the katabole encompasses both the act of expulsion and the excess that is expelled.

Bataille had a special interest in the katabole, seeing in it the potential for a reversal of scientific transparency and a recuperation of solarity as a light from below. Under the influence of this concept, he deploys the mythological images of the solar anus, the Jesuve, the pineal eye, and others, not to celebrate solar light, but rather to cast solarity itself into a carnal planetary morass from which it could recover and discover anew its own disavowed mythology. For Bataille, science had cast away not just the mythological but also the philosophical. It had ruthlessly divided consciousness in order to pursue a pure transparency. He thus wages his critique in the form of a mythological representation whose operation would present to science what it had violently expelled (94). Solarity appears as the antithesis of the illumination of consciousness, a reversal of the katabolic movement of the expulsion of myth through the application of the sun's destructuring effects on the subject and on consciousness itself. Thus, when solarity is figured in and by the mythological image, it returns to scientific discourse as a profane stranger that goes unrecognized by the consciousness that cast it out and disfigured it by expelling it in the first place. Upon its return to consciousness, the mythological image katabolizes the concept from which it was violently separated. It awakens phantasms within the very language of the concept.

The phantasm relates to the absolute concept as an elemental phenomenon. As Gasché (2012: 157) points out, Bataille's phantasms are *cold* and even have an icy effect on the topography of the mind. We might consider this passage from his essay "The Pineal Eye": "For if the affective violence of human intelligence is projected like a specter across the deserted night of the absolute or of science, it does not follow that this specter has nothing in common with the night in which its brilliance becomes glacial," (Bataille 1985: 81). It is for this reason Gasché (2012: 157) characterizes the violence of Bataille's mythological images and their heterogeneous effects as "glaçant," a quality that suggests at once the frozen environment of origin from which the phantasm emerges, an icy ghost that arises from the shadow cast by the absolutism of science, but also an image that mirrors (like *glace* or ice) the process of philosophical illumination as a sharply cold analytical mode.

In order to expose, mirror, and reverse the implicit violence of science and philosophy, Bataille mimics the method and disposition of objective authority. "The Pineal Eye" reenacts Hegel's primal scene at the origin of abstraction, when the Absolute Idea illuminates and claims images that fly up from the nocturnal pit of consciousness. In Hegel's (1983: 87) *Philosophy of Mind*, the nocturnal pit is an infinite world of images stored in the treasury of the Spirit, in the night of the human being. The nocturnal pit is a kind of automated and arbitrary geyser of sorts, shooting out images from its store in an archaic, nonsensical form. The Absolute Idea slumbers here, but begins to emerge through a "recollection" of the image, a procedure of illumination, appropriation, and refinement of the image that it claims from its archaic origin in the night. The Absolute Idea imagines itself into being through its claiming of the image. It appears out of its own sign-making fantasy, in which the illumination of the image in the night of being is the fundamental operation of abstraction.

Bataille reads Hegel's origin of consciousness as fundamentally *appropriative*, which he counters with a reverse operation: philosophy as expenditure. His heterology entails the individual subject of consciousness, expelling its contents *as its form of gratification and expansion*. Moreover, those unassimilable elements—the excreted phantasms—would be constituent actants of the very sociality with which Bataille counteracts the abstractions of science and philosophy. In this way, Bataille reverses Hegel's procedure of abstraction by casting an incandescent illumination into the nocturnal pit of the imagination, leaning further into the depths of consciousness to draw out and unleash its phantasms. Again, Bataille situates illumination as coming from the depths. This is a paradoxical light that returns from the nocturnal pit to cast itself on absolute concepts and reveal the phantasms that cling to them like cold residues. Solarity is this very excessive illumination: an elemental condition by which the phantasm freezes and sticks to the absolute concept and refuses to free itself. Gasché (2012: 157) explains: "Bataille does not oppose the warmth of affect and violence to the coldness of the abstract and the Absolute. On the contrary, what irrupts within it is a frost even icier than the coldness of understanding and reason. The frost of this affective violence congeals the coldness of reason."

Bataille releases elemental disfigurations to intensify analytic rigor. The play of planetary forces on concepts restages disciplinarity itself as a sadistic force that shatters foundational concepts. The sun appears as a disfiguration of the faculty of judgment writ large. It is unimaginably cold, not only in the sense that it demands a particular sacrifice (and does not care who that particular is), but because its exposure of the planet is totalizing. It exposes universally but demands a splitting into infinite singularities. Further, it demands its sacrifice as the cyclical and perpetual donation of singularities. Its coldness is the predatory impulse of consciousness: it hunts from the depths, and possesses us to do its hunting for it. The torment of Bataille's heterology lies in this contradiction: that it imagines the coldness of the sun as a form of blindness to the pain of ontological solitude (the sacrifice, the accursed share), and positions this blindness as the pretense of economic exchange.

Hot Planet: Extreme Incandescence In the Light of Cold Facts

In an especially difficult section of "The Pineal Eye" subtitled "The Sacrifice of the Gibbon," Bataille orchestrates another version of the violent exchange between the sun and the earth, this time with a dreamlike sequence featuring a group of nude men who torture a female gibbon and sacrifice her to the sun seemingly at the bidding of a perversely sexually aroused (white, bourgeois) Englishwoman. The group ties the gibbon to a stake—"trussed up like a chicken"—bury her alive in a hole in the earth leaving only her anus exposed (Bataille 1985: 85). Then they burn the skin of her protruding anus. She screams and her body convulses, seemingly for the gratification of the Englishwoman who trembles in synchrony with the gibbon's agony. The words are painful to read. Taken together, the entire assemblage, a sacrificed primate, the profane ecstasy of the Englishwoman, and the somnambulistic men who carry out the sacrifice, constitutes for Bataille a vision of economy under the cold light of the sun as the "tender pact between belly and nature" (85). He even sets the stage for this passage with reference to the "solar significance" of the feeling of pride and triumph of "the man perceiving his own dejecta under the open sky" (85).

Bataille conjures phantasmatic and phantasmological solar societies. These societies are phantasms that serve as their own analytical force; the afterburn of a drive to know the world by violently territorializing it and driving the planet to mass extinction. In Bataille's time, the nuclear bomb was the threat to be defused, but today nuclear energy has been accompanied and overtaken by the regime of global petroculture. The sacrifice of the gibbon might be thought as a precursor to the mass extinction of animals and other forms of life due to climate change. The planet is as hot as ever, not just in the earthly sense of overflowing volcanoes as Bataille writes, but now as a pervasive ecological condition: warming waters, air, wildfires.

We might therefore consider solarity as a contemporary planetary phenomenon: a disfiguring force that preys on us in and through the heat of the planet. In these terms, it is both the cause and effect of the petrocultural regime. It could very well mutate further as we transition to other forms of energy unless we understand, as Bataille does, that solarity is generated at the axis of the sun and the earth. It may be petroculture, with its unrelenting drive to yield the earth as an energy, that appears as the culmination of solarity, and not solar energy cultures at all. How else to explain the appearance of the Jesuve-like President of the United States, Donald Trump, whose very platform was a return to coal energy as a way of taking the reins in a global economy dominated by oil? Or how to explain the unrelenting aggression animating inter-ontological relations (between people across borders, against animals and other species)? Can it even be called an ontological relation if the boundary between beings is so devastatingly violated that the heteronormative white man's expenditure is the most dominant planetary force? The rise of climate change and the growing consciousness of the entanglement of global politics and planetary cataclysms is a realization of Bataille's perverse and paradoxical elemental intimacy. It is as though Bataille's phantasmology has been accepted as the normative operation of life itself, as the privileged form of biopower, as a force that disciplines life by sadistically exerting itself against, expelling itself into and out of living things not only to their death but to their very extinction. Nevertheless, it is this operation that has overtaken the (non)relationship between humans and other living species, hypostatized as racial and gender difference, and dramatized by political exchange between nation states. The exposure and consciousness of life and lives has never been colder, and the planet has never been hotter.

Yet it is here that we must reflect on Bataille's aesthetic maneuver: his disfigurations burn out the figure, and they reveal that the figure was always destined by the depths in which the sun burns. If we took hold of our solarity, as Bataille invites us to, and we were to look coldly at our economy, its epistemic underpinnings, and its implicit demands for sacrifice, what would we find? What would it mean for us to come to terms with the lens of our solarity today? Would the energic underpinnings of the global economy seem more real or more fantastical? More torturous or more factual? Further, what would it mean to embrace the lens of solarity, rather than blindly let it (continue to) power our culture and our future? For inasmuch as Bataille forces us to dwell with the pain of sacrifice it is precisely because of his interest that such a figurative maneuver provokes deep reflexes—revulsion surely—but also, as Allen Stoekl suggests, the recoil of transgression is the unexpected site of ethics (Stoekl 2007: 252–282). The phantasms that animate Bataille's writing cannot but be seen as despicable, once they are seen. The Jesuve, the lovers, the Englishwoman, the priest, and all the figures that populate Bataille's essays derive their solarity from the sheer misery of degradation and torture they exert. Indeed, their pleasures are precisely the fulfilment of their own excremental mode. Bataille binds these phantasms to their imperial, bourgeois, white, and normative absolutes so that we locate their origin in the appropriative procedures of concept formation in the Enlightenment tradition. Might we, then, accept the challenge Bataille presents to us and take hold of solarity by expending it?

References

Baudrillard, Jean. 1998. "When Bataille Attacked the Metaphysical Principle of Economy." In *Bataille: A Critical Reader*, edited by Fred Botting and Scott Wilson, 191–95. London: Blackwell.

Bataille, Georges. 1985. *Visions of Excess. Selected Writings, 1927–1939*, edited by Allan Stoekl. Minneapolis: University of Minnesota Press.

Bataille, Georges. 1991. *The Accursed Share: An Essay on General Economy. Volume 1 Consumption*, translated by Robert Hurley. New York: Zone Books.

Gasché, Rodolphe. 2012. *Georges Bataille: Phenomenology and Phantasmology*, translated by Roland Végsö. Stanford: Stanford University Press.

Haraway, Donna. 2016. *Staying with the Trouble: Making Kin in the Chthulucene*. Durham, NC: Duke University Press Books.

Hegel, Georg Wilhelm Friedrich. 1983. *Hegel and the Human Spirit: A Translation of the Jena Lectures on the Philosophy of Spirit (1805–6)*. Detroit: Wayne State University Press.

Latour, Bruno. 2004. *Politics of Nature: How to Bring the Sciences into Democracy*, translated by Catherine Porter. Cambridge, MA: Harvard University Press.

Spivak, Gayatri. 2003. *Death of a Discipline*. New York: Columbia University Press.

Stengers, Isabel. 2015. *In Catastrophic Times: Resisting the Coming Barbarism*. London: Open Humanities Press.

Stoekl, Allan. 2007. "Excess and Depletion: Bataille's Surprisingly Ethical Model of Expenditure," *Reading Bataille Now*, edited by Shannon Winnubst. Bloomington: Indiana University Press.

Daniel A. Barber

Active Passive:
Heat Storage and the Solar Imaginary

There is no such thing as *the* house, or the house as such, there are only historically and culturally contingent cultural techniques of shielding oneself and processing the distinction between inside and outside.
–Bernhard Siegert, *Cultural Techniques: Grids, Filters, Doors, and other Articulations of the Real* (2015)

In what follows, I aim to justify an inversion or least a productive complication of the familiar nomenclature of active and passive solar energy, as it pertains to architectural design methods and to solarity more generally: that is, to changes in economies, cultures, and ways of living in the present and future. In the received active/passive distinction, *active* denotes a solar energy system that requires a mechanical device of some sort, usually photovoltaics that convert solar radiation into electricity. Such systems aim to replace fossil-based forms of electricity generation and heat provision. Think, for example, of Tesla's Solar Roof: photovoltaic shingles that can be attached to virtually any roof design and plugged into the house system (often accompanied by a Tesla Powerwall battery). Over the past few decades, research into solar energy in the architectural context has tended to focus on just such technology-intensive approaches

The South Atlantic Quarterly 120:1, January 2021
DOI 10.1215/00382876-8795754   © 2021 Duke University Press

towards solar and insulative efficiency. Passive systems, by contrast, rely on design rather than technological innovation. They involve attention to orientation and site, thermal properties of materials, seasonal shading devices, and other means to selectively absorb and optimize the heat or light directly from the sun. Generally speaking, passive systems don't require additional resource input, or minimize it–perhaps just a fan to distribute sensible heat. Though passive on these terms, such houses often require the inhabitant to engage with them in an elaborate fashion, by drawing insulating curtains at night, focusing time in one part of the house or another according to season or time of day, and modifying expectations of the thermal interior so that consistency is less important than efficiency.

Active solar, in this received framework, has no, or very few, design implications. Because of this, much of the professional and popular discussion of solar energy, in the context of buildings, involves efficiency, placement, and material options for active, photovoltaic systems. Passive solar, by contrast, is all design: the house, the patterns of the inhabitant, and often a range of thermally induced social practices are solicited by and through such projects, adjusting existing habits and ways of living in interior spaces, in order to allow for a new kind of social interaction with the sun. The house, its capacity to absorb and store heat, in concert with activities of the inhabitant, articulate the system: solar energy, in this framework, is an interactive medium, reliant on participation, articulating a social relationship to fuels appropriate for a given condition.

At stake in reconsidering this distinction is an attempt to understand how we simultaneously experience the resource conditions of our thermal interiors and the transformations of global climatic patterns. Which is to say, to reconsider active and passive in solar architecture (with heat storage, as will become clear below, as the hinge) is also to reimagine the role of buildings in the production of the carbon zero future—less, at least relatively, as spaces of technological innovation, and more as spaces of social and species evolution. An *active passive* solar architecture aspires to lifestyles, habits, and expectations coming into line with the massive geophysical transformation of climate instability. The house becomes a medium for this epochal social change. In addition to exploring photovoltaic-induced technical efficiencies and allegiance to the parameters of performance software, sensor data, and adaptive comfort (that is, broadly construed, the technical fix of the ecomodernists), *active passive* solar architecture also opens up a space for exploring radical changes to means of global inhabitation (interior and exterior). Architecture in this fashion can be framed as a medium for the Anthropocenic transformations solicited by climate instability.

At stake as well is reconsidering the basic economic model for energy systems that encourage specific ways of living on the planet. Unlike buried fossil fuel reserves, which are subject to analysis relative to cost of extraction and time to exhaustion, passive solar technologies are evaluated for their relative efficiency of radiation collection and storage, and subject to increases in that efficiency as technology improves. As Eugene Ayres, a Gulf Oil executive sympathetic to a solar energy transition, put it in 1948:

> The most important factor is not the size of a reserve, but the rate at which it can be procured. When the rate-based measurement is applied to a process that is essentially unrenewable, such as the combustion of fossil fuels . . . the answer is related to the number of years during which that source could be relied upon. On the contrary, if the source is relatively continuous . . . the answer is the rate at which power might be obtained for the almost indefinite future. The two kinds of figures are not comparable. (Ayres and Scarlott 1952: 12)

This economic incompatibility of solar energy and fossil fuels can be emphasized and celebrated as an opportunity for new forms of collective energy management, in addition to being a spur towards specific kinds of design and lifestyle innovation. Solar-engaged architecture is an effective means towards facilitating novel and regionally appropriate socio-economic arrangements. Again, new frameworks for design have been essential to these articulations, as have explorations of heat storage, activating the inhabitant as an agent in increasing the "rate at which power might be obtained" through often straightforward habits and practices.

Along with the solar panel, insulation, and the volumetric disposition of the interior, heat storage is an essential element in the design of the solar house. "The problem of solar energy," as Maria Telkes (1947: 12), one of the pioneering engineers in solar energy, clarified, "is largely a problem of heat storage." It is relatively easy to design a house to absorb the sun, but much more difficult to enact a storage regime that allows solar energy to heat a house through the night, or for days without sunlight. Of the elements of solar house design, heat storage emphasizes practice and process—not just a determinant of the design itself, but of how the design proposes new ways of life in the interior. Whereas solar energy may invoke a sense of abundance and abstract value, storage is the operand of limitation. It holds solarity in check.

If the house (or residence more generally) is one of the primary media through which social bodies articulate a relationship to fuels (i.e., articulate a concept of 'energy' in its most basic sense), then heat storage is the medium that allows a solar house to produce a new sort of subject, a solarized inhabi-

Figure 1. George O. Löf and James Hunter, Löf House, Denver, Colorado, USA, 1957.

tant, ready and able to adjust patterns, lifestyles, and aspirations through meticulous engagement with solarity (Daggett 2019: 3).

Histories of heat storage in this sense premediate the built interior as a dynamic space of social contestation: of the connection, now seemingly inevitable, between the mechanical (fossil-fueled) conditioning of the interior and the emissive (fossil-fueled) conditioning of the atmosphere. In order to play out the premise of the *active passive* solar house, I will explore and contextualize three significant entries in the history of residential heat storage. Geographically, these examples focus on North America in the context of broader global trends; historically, they are bracketed by World War II on one end, and the search for oil and other resources that the end of the war instigated, and the end of the second oil crisis in the 1970s (c. 1978) at the other end, a moment when the initial excitement about renewables in general and the solar house in particular began to wane. In each of these case studies, I aim to clarify the historical value of these buildings and also their promise, both materially and symbolically, for opening up new avenues for building, thinking, and living in solarity.

Löf House, Denver, CO, 1957

The George O. Löf House was built in Denver, Colorado, completed in 1957 (Figure 1). Löf was an engineer and collaborated with the Boulder-based

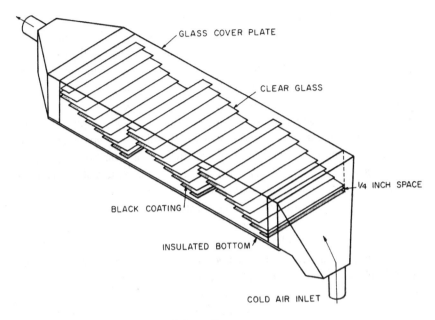

GLASS COVER PLATE

CLEAR GLASS

1/4 INCH SPACE

BLACK COATING

INSULATED BOTTOM

COLD AIR INLET

Figure 2. George O. Löf, overlapped plate collector, c. 1947.

architect James Hunter on the design of the house. The low, narrow build-ing reflected what was, by then, a well understood set of principles in design-ing for solar heating: an open, glazed southern exposure, ideally of dou-ble-paned glass panels to allow for some insulation; a carefully determined roof extension that shaded the interior or let the sun it in according to sea-sonal patterns; and a narrow, compact design that allowed for most of the living spaces to be subject to these seasonal adjustments.

The Löf House was a living experiment in solar heating. The roof was flat, with solar collectors placed atop a plinth. The collectors were indepen-dent from the flat roof and angled for optimal absorption of radiation. Air cir-culated through these collectors (as opposed to water or another medium), and this heated air was then drawn by fans into the storage system. Löf deployed what he called the "overlapped plate collector": layered, overlapping glass plates, selectively painted with a black coating to maximize solar absorp-tion, and encased in a sealed and insulated panel. These collectors generated relatively high heat conditions in relationship to the air outside (Figure 2).

Once heated, the air was circulated by fans to "heat storage beds" filled with river pebbles. These "beds" took the form of columns extending down the middle of the house's central stairway: as the heated air entered the columns, the pebbles would warm up and store the heat; as the air in the house (around the columns) cooled, the pebbles would release their heat and it would radiate into the house. There was also a mechanism to move this heated air through the building by means of forced air vents, familiar to other forms of furnace-based hot air heating systems (and complicating the passive denotation—some additional energy input was required to optimize the system).

The heat storage beds followed on a range of ongoing experiments in heat storage in the years immediately following World War II, which saw extensive experimentation in the design and technology of solar heating (before photovoltaics) (Barber 2016: 63ff). The most common medium of heat storage was water—circulated through solar collectors, the water would be heated and stored in an insulated tank. When house heat was needed, air could be blown over the warmed water and into the house. Löf's first such experiment, in a small house in Boulder, Colorado built in the 1910s (i.e., not designed as a solar house but retrofit) also used an overlapped plate collector and stored the heat in a pile of pebbles in the basement; the heat could be drawn into the interior as needed, as an adjunct to the house's existing furnace.

At Löf's Denver house, the columns were intended to be "worked into the decorative scheme of the house" and thus an essential aspect of the design—revealing rather than hiding the energy infrastructure (Figure 3). In order to make the house work, an elaborate regime was proposed, "a new living pattern," as Hunter, Löf's architect, put it: the occupants (the Löf family, dedicated to the promise of solar living), would be "happy to draw insulating curtains, adjust controlling louvers, and assist the sun in performing the tasks we have assigned to it" (Hunter 1956: 205). Around the same time, Hunter was the architectural consultant for an international competition to design a solar house. In the brief distributed to interested designers, he spoke of the potential occupants of the house as having "great respect for the sun and its influence on their way . . . they are adventuresome enough to commit themselves to this form of energy" (Hunter 1958: ii).

The Löf House was, technically, a hybrid system, relying both on the radiation of heat from surfaces and on the use of electrical fans and other adjuncts. It was similar to a number of experimental houses built by the Chemical Engineering Department at the Massachusetts Institute of Technology between the late 1930s and the early 1970s—Löf worked on the first of those houses while a student in the department. Like many of his colleagues, he was disappointed with the results and sought to innovate in both collector efficiency and storage possibilities. Maria Telkes, a researcher in the department in this period, quoted above, developed a method of heat storage using the phase-change capacity of chemical solutions, which would

Figure 3. Columns filled with heat-storing river pebbles, in the Löf House. Photograph courtesy Anthony Denzer.

solidify as they absorbed heat received from a panel, and then, as the air around them cooled, these chemical solutions would liquefy and radiate heat to the interior. Telkes used what she termed "heat bins," an analogue of Löf's heat beds: material insertions into the design of the house that stored collected heat for distribution as needed. Her system didn't work very well, though it initiated a research trajectory that has since borne significant fruit in the elaboration of phase-change materials (PCMs): standard building materials, such as sheetrock, in which such chemicals are embedded in order to give the material some capacity to store and later radiate heat.

At the Löf House, the equation was this: in order to articulate a viable solar heating system, one needed not only a carefully designed house, an effective solar panel, and an integrated storage system, but also a willing participant. Indeed, the role of the inhabitant was essential to activating the possibilities of solarity, in large part though managing various elements to optimize heat storage capacity. Importantly, especially for houses in cooler climates, this also implied a willingness to adjust one's expectations: to wear a sweater, for example, or focus time in warmer rooms. Solar energy for house heating required the willing participation of the inhabitant: it required activity.

Kelbaugh House, Princeton, NJ, 1974–75

The Kelbaugh House was designed and built by Douglass Kelbaugh in Princeton, New Jersey in 1974, after he graduated from the Princeton School

Figure 4. Douglass Kelbaugh, architect. Exterior view of Kelbaugh House, Princeton, New Jersey, USA, ca. 1975. Chromogenic color print, 20.5 x 24.5 cm, ARCH254206 Douglas Kelbaugh Fonds, Canadian Center for Architecture. Gift of Douglas Kelbaugh.

of Architecture in 1972 (Figure 4). It used what had become known as a Trombe Wall, alongside a range of insulation and ventilation techniques and living habits, to keep the house warm in the frigid Northeastern winter and to maintain comfort in the hot, humid summer.

The Trombe Wall, as a concept and a practice, was by this time well known. It follows a basic principle somewhat at odds with the open living of the house described above: in most post-war passive solar houses, the sun-exposed façade was filled with glass, both to bring the sun into the house and celebrate the solar lifestyle. The Trombe Wall instead takes advantage of solar radiation to maximize potentials of thermal mass: the sun-exposed wall has no (or few) windows, and consists of a thick membrane of a thermally dynamic material (usually concrete or masonry) that absorbs the radiation during the day, and distributes it to the interior at night. The principle is named after the French engineer Félix Trombe who, working with the architect Patrick Michel, designed a small experimental house in the Pyrenees with a wall of thick concrete in 1967 (Figure 5).

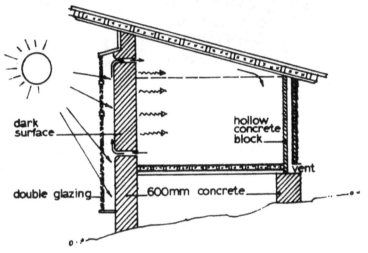

Figure 5. Felix Trombe and Patric Michel, Trombe House, near Odeillo, France, 1967. Photograph and sectional diagram.

The south façade at the Kelbaugh House is a 15-inch-thick concrete wall, set directly behind two panes of "double-strength window glass"; the wall was painted in a black "that absorbs more energy than it emits back." The concrete wall absorbs radiation, the double-paned glass sandwich holds that radiation in, and a series of flues, vents, and interior openings distribute the heated air around the house as needed. The sectional drawing uses the graphic strategy of arrows of varying thickness and trajectories to suggest the dynamism of the system (Figure 6). As Kelbaugh described it, "The warm wall and glass heat up the air which then rises up the slot"; the heated air can

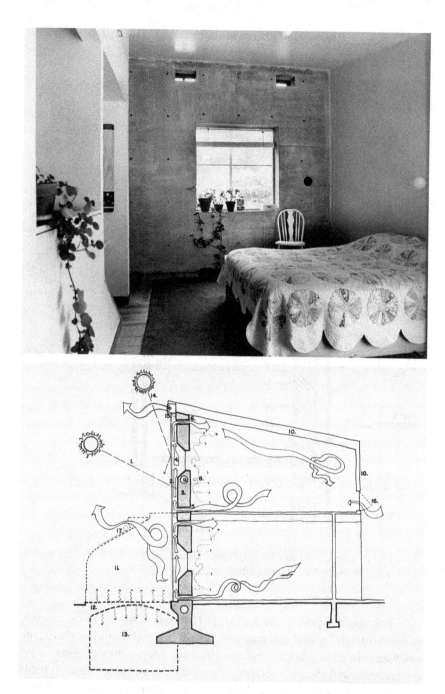

Figure 6. Douglas Kelbaugh, architect, Interior view of the Kelbaugh House, Princeton, New Jersey, USA, ca. 1975. Gelatin silver print, 20.5 x 25.5 cm, ARCH283920 Douglas Kelbaugh Fonds, Canadian Center for Architecture. Gift of Douglas Kelbaugh; and section showing the circulation by thermosphere for the Kelbaugh House, Princeton, New Jersey, USA, ca. 1975. Electrophotographic print on paper, 21.6 x 28.1 cm, ARCH254191 Douglas Kelbaugh Fonds, Canadian Center for Architecture. Gift of Douglas Kelbaugh, ©Douglas Kelbaugh.

be ventilated in the summer; in the winter, it is drawn into the house and heats the rooms—first the spaces closer to the wall and then the northern side of the house. The rising, heated air also pulls up more cold air into the system, allowing the process to continue. The house was heavily insulated so that a relatively small amount of radiation could keep it warm at most times of the year; there was also a small backup furnace system. The heating system also involved the greenhouse and the small cellar as heat storage.

The Kelbaugh house was but one example of the dynamic proliferation of the Trombe Wall and numerous other means of "passive" heat storage in the 1970s. Indeed, in the midst of that decade's oil scares, experiments proliferated. Alongside the Kelbaugh House, the best known is likely Steve Baer's house near Corrales, New Mexico, built in 1971 (Figure 7). The sun-exposed wall consisted of a gridded scaffold populated with ten-gallon drums filled with water. This scaffold sat barely inside the house, behind a protective, insulated metal-framed wall on a hinge: in the day time, the wall would sit flat in the front of the house, allowing solar radiation to heat the water in the drums; in the evening, the wall would be cranked up to seal the house, thereby providing a thick insulative membrane. The heat stored in the drums would radiate into the interior as the air around them cooled. Another example: the Aspen Airport Terminal designed by Copland, Sinholm, Hagman, Yaw Architects in 1975, in Aspen, Colorado, used a "Bead Wall" system (Figure 8). The south façade consisted of two thick layers of glass with an air cavity between. During the day, solar energy could flow into the interior; at night, the air cavity would be filled with Styrofoam beads in order to insulate the building and store the collected heat. As the sun rose, a vacuum would draw the beads back out, opening the building to radiation once more.

The period provided a number of opportunities for elaborations on and adjustments to formal and technological standards in the built environment, as a means towards maximizing energy efficiency. These practices were documented initially in journals, and very soon afterwards—still in the mid-1970s—in a number of government, university, and industry sponsored manuals. In 1975, a collaboration of firms in Harrisville, New Hampshire, under the name of Total Environmental Action produced the book *Solar Energy Housing Design in Four Climates* for the Research Corporation of the American Institute of Architects, under a grant funded by the National Bureau of Statistics. It was filled with schematic descriptions of different solar building types, lengthy indexes outlining recommendations for ideal diversion from a due south building orientation, as well as an index of precise tilt-angles, according to latitude and elevation, to maximize efficiency of solar collectors. A number of solar heating systems were described with

Figure 7. Steve Baer, Baer House, near Corrales, New Mexico, USA, 1971.

technical diagrams, more charts showing the precise relationship of the solar collector area to the cubic volume to be heated, and with plan drawings showing the most amenable distribution of interior rooms and programs according to solar conditions. Site plans and general climatic analyses (i.e., also including wind, precipitation, humidity) also accompanied these drawings, intending to provide a comprehensive approach for other sites and design challenges (Total Environmental Action 1975: 23). The catalogue was developed out of another book produced by the group in 1973 on "solar shelters;" Total Environmental Action also published a guide to "natural thermal heat storage" with the Department of Energy in 1979.

A collaboration between the AIA Research Corporation and the College of Architecture at Arizona State University (ASU) was one of the first attempts to comprehensively map the solar types appropriate to different solar regions, through documenting and analyzing built projects in each region. The Project Director, John I. Yellott, had been the Executive Director of the Association for Applied Solar Energy, headquartered at ASU, when it ran an international competition for a solar house in 1958 (the competition for which Löf's architect James Hunter was a consultant). In addition to general technical and design parameters, according to region, the book mostly consisted of examples laid out in identical format with an information sheet and a hand drawn perspective (Figure 9). Dates and location, participating designers and engineers, general climatic data, building size, and materials framed a precise discussion of collection area, heat storage system, and auxiliary systems. Relevant publications or notes from research team visits were

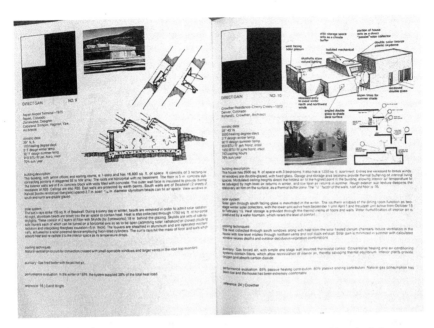

Figure 8. Copland, Sinholm, Hagman, Yaw Architects, Aspen Airport Terminal. Aspen, Colorado, USA, 1975.

also included. A final section of the book analyzed selected projects in more detail and proposed a rubric for best practices (Yellott 1975: 56ff).

A similar publication came out in 1979 from the AIA Research Group, the Office of Policy Development and Research and the Division of Energy, Building Technology, and Standards of the US Department of Housing and Urban Development, and the US Department of Energy. *A Survey of Passive Solar Buildings* was, again, organized according to both type and region; the book used photographs instead of axonometric drawings and focused its evaluation of the buildings it catalogued according to their performance since being built. Directly reflecting Telkes's 1947 intervention, the projects were indexed according to heat storage system, according to the following categories: Sunspace, Direct Gain, Roof Pond, Trombe Wall, Water Trombe Wall, and others (AIA Research Corporation 1978: 21ff) (Figure 10).

Saskatchewan Conservation House, 1977

A third house shifts the discussion. Designed by a committee of research-ers led by Harold Orr and built in 1977 as part of the research efforts of the

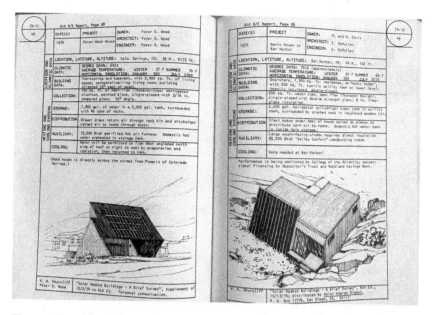

Figure 9. Spread from John I. Yellott, *Solar Oriented Architecture: 1975 Research Report.*

University of Saskatchewan, the Saskatchewan Conservation House relies less on precise methods of heat storage and more on the house itself as a super-insulated storage system. Insulation for the house was six to ten times that of typical construction in the period (Orr 2015) (Figure 11). The research team, having determined that the frigid northern winters did not justify solar heating, looked instead, as the "conservation" appellation suggest, at means of holding in the relatively small amount of radiation that was absorbed.

The nuclear physicist and solar housing advocate William Shurcliff celebrated the house's "thick, clever, and thorough insulation," as he began to advocate for insulation as the key factor to a successful renewable-energy based building in the late 1970s. In his 1981 book *Super Insulated Houses and Double Envelope Houses: A Survey of Principles and Practices,* he argued for the promise of a house that "is so well insulated, and is so airtight, that, throughout the winter, it is kept warm solely by (a) the modest amount of solar energy received through windows and (b) miscellaneous within-house heat sources;" the latter he lists as: "stoves for cooking, domestic hot water systems, clothes dryers, electric lights, clothes washing machines, dish-washing machines, human bodies, TV and radio sets, refrigerators, etc." (Shurcliff 1981: 44).

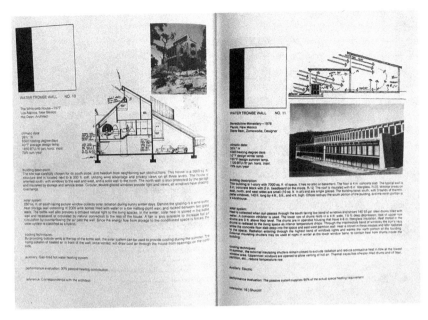

Figure 10. Spread from *A Survey of Passive Solar Buildings*, 1979.

Shurcliff was dismissive of much of the solar house activity of the 1970s glossed above; in a number of publications, such as *Solar Heated Buildings of North America: 120 Outstanding Examples,* he documented, analyzed, and critiqued what he saw as scattershot attempts to refine solar technologies (Shurcliff 1978: 13). He later told the *New York Times* that he published the book because trying to promote solar design "was infuriating . . . [in any given example] half the information was missing, and systems were vaguely described as 'ingenious' without explaining how they were ingenious or how well they worked. Sense had to be made of it" (Shurcliff 1980b). Rather than regulation, information could circulate, he proposed, as a means to improve methods and techniques.

The basic premise of super-insulation was later developed through experiments then ongoing in a number of departments of the emerging field of architectural science—in Saskatchewan, also at the University of Delaware, the University of Sydney, the Technical University of Denmark, and through a government funded research consortium in Aachen, Germany. In 1990, the German physicist Wolfgang Feist codified some of the knowledge and principles gained from these experiments in the form of the now well-known *Passivhaus* system (Feist et al. 2005).

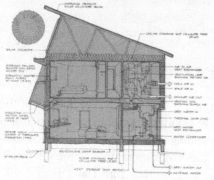

Figure 11. Saskatchewan Conservation House, near Regina, Saskatchewan, 1977; photograph of house and sectional diagram.

The emergence of super-insulation opens up two seemingly contradictory trajectories. On the one hand, the design guidelines that surround the passive house system suggest, indeed, a sort of passive engagement that eschews the active habitation that the previous two houses elicit. Once well designed, the implicit narrative goes, a passive house will not require extensive quotidian engagement from the occupant—in this regard, it follows the familiar path of other heating and cooling strategies. Tightly sealed against the elements, it relies on technical efficiencies rather than social practices to maximize its energy saving potential.

Yet, returning to Shurcliff in the late 1970s, a political position was embedded in this super insulative approach. Shurcliff's extensive review of solar house design ideas over the decade, while at times dismissive, other times celebrated the variety and creativity exhibited by the range of approaches. Shurcliff's 1976 article "The Case against Government Standards for the Solar Industry" aimed to substantiate the need for the *unregulated* elaboration of solar energy technologies, in the face of increasing governmental attempts to manage the solar energy market. In the US, the 1974 Solar Heating and Cooling Demonstration Act and the Solar Energy Research, Development and Demonstration Act regulated solar research and required the certification of solar energy systems. These were seen as intrusions on the free experimentation that was crucial to the solar design process; as Shurcliff (1980a: 26) put it, "mandatory criteria [established by the government] would tend to inhibit invention." Shurcliff's frustration reframes the active passive dynamic, suggesting a tether between the activities of the solar heated interior and broader political conditions and constellations.

Contingency and Solarity

The built world is contingent—every structure we inhabit was built according to a set of principles and priorities; every structure can be retrofit, or built again. Beyond the practicalities of solar living, the contingency of the built environment reflects back on a more general interrogation of the historical necessity of contemporary ways of life. As Andreas Malm (2016: 13), in his history of fossil capital, suggests with a potent building metaphor:

> Agents must have created [the fossil economy] through events amounting to a moment of construction, much as, once erected, a building's structure is now an enduring feature in the world; entrenched in the environment, it conditions the movements of the people inside. Eventually it appears indistinguishable from life itself: business as usual. But the fossil economy was once constructed and has since been reproduced and enlarged, and anything built over time can potentially be torn down (or escaped).

The history of solar house design is thus a past that has potent resonance with the future—for the future of architecture, to be sure, and also for how, on more general terms, we imagine, speculate upon, and design for life amidst increasing climate instability. *Active* systems are from this perspective *passive*: they assume a technological fix, with no demands on and little change to the social body—again, the solar roof tile simply replaces previous roof tiles, with no adjustments to the life going on in the house. *Passive* systems, on the other hand, require novel material engagements and focused attention to solar patterns, geophysical systems, and future possibilities relative to climatic and other conditions. They insist on the importance of individual and collective agency. In the *active passive* house, the epistemological framework is less one of the performance of technical efficiencies, and more one of how the habits, practices, and social values performed on the interior simultaneously reflect and induce new possibilities of zero carbon living.

This inversion has compelling resonance across the broader socio-political challenges of solarity. Solar energy has long been seen as a technology imbued with inherently democratic potentials: as Langdon Winner famously noted in 1980, advocates of the 1970s saw solar as desirable "not only for its economic and environmental benefits, but also for the salutary institutions it is likely to permit in other areas of public life" (21). Active solar energy, in the form of roof tiles indistinguishable from any other roof tile, seems to resist this general political premise; that is, the application of photovoltaics is more of a continuation of status quo patterns of extraction politics, corporate energy management, and the eco-modernist technical fix than it is a reimagining of social institutions towards new ways of life. Although the single-family houses described here fall far short of any political program or model for alternative activities, they nonetheless gesture, at the least, towards a novel disposition. Towards, in other words, eliciting a solarity that frames cultural transformation—active engagement with the production of new ways of life, often through design—as the site for exploration and research into the solar future.

References

AIA Research Corporation. 1978. *A Survey of Passive Solar Buildings*. Washington, DC: American Institute of Architects Research Corporation.

Ayres, Eugene, and Charles A. Scarlott. 1952. *Energy Sources: The Wealth of the World*. New York: McGraw Hill.

Barber, Daniel A. 2016. *A House in the Sun: Modern Architecture and Solar Energy in the Cold War*. New York: Oxford University Press.

Dagget, Care New. 2019. *The Birth of Energy: Fossil Fuels, Thermodynamics, and the Politics of Work*. Durham, NC: Duke University Press.

Feist, Wolfgang, Jürgen Schnieders, Viktor Dorer, and Anne Haas. 2005. "Re-Inventing Air Heating: Convenience and Comfort within the Frame of the Passive House Concept." *Energy and Buildings* 37, no. 11: 1186–1203.

Hunter, James. 1956. "The Architectural Problem." In *Proceedings of the World Symposium on Applied Solar Energy*. Menlo Park, CA: Stanford Research Institute.

Hunter, James. 1958. "Preface." In *Living with the Sun: Volume I: Sixty Plans Selected from the Entries in the 1957 International Architectural Competition to Design a Solar-Heated Residence*. Phoenix: Association for Applied Solar Energy.

Malm, Andreas. 2016. *Fossil Capital: The Rise of Steam Power and the Roots of Global Warming*. New York: Verso.

Orr, Harold. 2015. "Saskatchewan Conservation House." *Passipedia: The Passive Hosue Resource*, March 12. passipedia.org/basics/the_passive_house_-_historical_review /poineer_award/saskatchewan_conservation_house.

Shurcliff, William. 1978. *Solar Heated Buildings of North America: 120 Outstanding Examples*. Baltimore, MD: Brick House.

Shurcliff, William. 1980a. "The Case against Government Standards for the Solar Industry." In *Notes from the Energy Underground* edited by Malcolm Wells, 26–28. New York: Van Nostrand Reinhold.

Shurcliff, William. 1980b. "A Pioneer Inventor Recalls the Dawn." *New York Times*, December 4. timesmachine.nytimes.com/timesmachine/1980/12/04/111320946.pdf?pdf _redirect=true&ip=0.

Shurcliff, William. 1981. *Super Insulated Houses and Double-Envelope Houses*. Baltimore, MD: Brick House.

Telkes, Maria. 1947. "Solar House Heating: A Problem of Heat Storage." *Heating and Ventilating* 44, no. 5: 12–17.

Total Environmental Action. 1975. *Solar Energy Housing Design in Four Climates*. Harrisville, New Hampshire: Total Environmental Action.

Winner, Langdon. 1980. "Do Artifacts Have Politics?" *Daedalus* 109, no. 1: 121–36.

Yellott, John I. 1975. *Solar Oriented Architecture: 1975 Research Report*. Tempe, AZ: Arizona State University.

Jamie Cross

Viral Solarity: Solar Humanitarianism and Infectious Disease

> Our best machines are made of sunshine;
> they are all light and clean . . .
> —Donna Haraway, "A Manifesto for Cyborgs" (1987)

What does solarity look like in an epidemic?

In the midst of the 2014 West African Ebola virus epidemic a Dutch solar company, WakaWaka, launched an online fundraising campaign. Under the slogan "You can't fight Ebola in the dark," the campaign invited people to purchase a solar powered handheld torch and, by doing so, pay for another to be packed in emergency kits bound for Liberia.

The campaign went viral. Wakawaka sold US$70,000 worth of products in less than six hours, with people paying for solar lamps that the company pledged to send to medical clinics and health care workers. This was the first time an off grid solar energy company operating in Africa had connected their technology to an emergency public health response at such scale or had sought to establish a causal link between brand name products and the ability of medical professionals to treat victims of a disease.

Over the next twelve months the company's campaign site attracted 14 million visitors, raising

The South Atlantic Quarterly 120:1, January 2021
DOI 10.1215/00382876-8795766   © 2021 Duke University Press

the profile of its work and attracting praise from journalists. A corporate profile published in the *Guardian* newspaper described the company as a powerful, positive example of business action on the sustainable development goals; "a consumer brand trying to harness the power of the high street to benefit low income communities around the world" (Balch 2014).

Contrast this with the *Guardian's* portrait of another energy company as it intervened in the Ebola epidemic. In 2014 the chief executive of transnational coal mining company, Peabody, sought to establish links between burning coal and the humanitarian response to Ebola. At an international conference, the chief executive championed coal powered electricity as the solution to the problem of how to distribute vaccines and extend public health care across Africa (Goldenberg 2015). A leaked PowerPoint slide led the *Guardian* to accuse the company of profiteering.

The Director of Columbia University's National Centre for Disaster Preparedness and White House adviser on the US response to Ebola told the newspaper, "Peabody has very specific and explicit corporate goals. I think this is a pretty far-fetched leap from a global crisis to try to justify the existence of a company that is interested in producing and selling coal. . . . I think it's an opportunistic attempt and somewhat desperate to relate corporate self-interest to a massive public health crisis" (Goldenberg 2015). Peabody's senior vice president responded with a letter accusing the *Guardian* of bias against fossil fuel industries (Svec 2015).

In both examples, corporate executives sought to establish a causal link between their commodities and the public health response to Ebola. But while Peabody's actions were reported as a blatant attempt by the fossil fuel industry to capitalise on a crisis, WakaWaka's attracted praise. Whilst the interventions of the coal company appeared as barely concealed self-interest, those of the solar company appeared to represent a purer expression of humanitarian sentiment. Where coal appeared morally bankrupt, heavy with the weight of its own corruption, solar appeared ethically untainted; all light and clean. Yet, as I explore in this article, the combination of solar power with care for distant others and the alleviation of suffering is no less entangled in the pursuit of commercial interest, and no less bound to carbon in a history of energy, capitalism, and humanitarianism, than coal or oil.

This dichotomy between the moral economy of fossil fuels and the moral economy of solar power captures a particular aspect of what the editors of this issue of *South Atlantic Quarterly* invite us to call "solarity." That is, the possibility that the energy of the sun might engender alternative energetic and socio-political futures, and different ways of being in relationship to each other than those that have characterised histories of fossil fuel extraction. Yet as the contributors to this issue emphasise, solarity takes many different forms, characterised by different relations between people, other species, and the sun; and, we might add, by different relations to the histories and legacies of carbon modernity.

So, what characterises the kind of solarity manifested by solar companies seeking to expand access to electricity in contexts of humanitarian emergency? In many respects, it maintains important continuities with forms of fossil fuelled humanitarianism. It implies empathy and compassion for others, rather than solidarity. It is designed around the sun's energy, insofar as this energy can be transduced by a machine, to power products and services. It is resolutely techno-optimistic or techno-utopian. It is not the antithesis of a market relation: *it is* a market relation. It is also, we might argue, the form of solarity that has come to be transmitted more rapidly and widely, than any other.

How are we to critically apprehend the relationship between economic and moral logics in this form of solarity? As we struggle to respond to the worldwide SARS-CoV-2 (COVID-19) pandemic and begin to envisage the role of solar energy in a green recovery, these questions are poised to become more relevant than ever. This article draws on long term ethnographic research in the global off grid solar industry. In what follows I revisit the interventions of one company in the Ebola crisis, and lay the grounds for an anthropology of humanitarianism and solar power.

Contagion, Compassion, Energy Markets

WakaWaka was established in 2012 by two Dutch green energy entrepreneurs, taking its name from the pop song of the South African World Cup by Columbian artist Shakira, and a Swahili phrase meaning "shine bright." During the 2010s it was one of tens of solar companies established in Europe and North America to manufacture and sell solar lighting and charging products in Sub Saharan Africa. Over this decade some one hundred eighty million products were sold by off-grid solar companies across Africa, Asia, and Latin America, thirty million of them alone between 2018 and 2019. By March 2020 the annual market for off grid solar energy products was valued at US$1.75 billion (World Bank 2020).

In 2014 the company's co-founders followed news stories about the Ebola epidemic from their headquarters in Amsterdam. "We heard that doctors were actually operating in the makeshift tents in the field, by the thousands, to help Ebola victims, and we realized that these were in the off-grid areas where we were working," one of them told me.

At a staff meeting to talk about future opportunities for marketing their products in Africa, they discussed how the company's mission—"to make an impact in developing countries"; "to use the energy behind our business to make the world a better place"; "to preserve the planet by developing products that respect the earth and the environment"—related to the Ebola epidemic. "The question just came up: how do we make the connection between light and Ebola? People are dying because of an infectious disease: what on earth has light got to do with it?"

The question catalyzed an exercise in "design thinking" (Rowe 1987; Gunn, Otto, and Smith 2013) that saw the company's employees try to put themselves in the shoes of Ebola victims and doctors. One of the company's founders talked me through the process when I interviewed them in 2018:

> We started to envision or visualise ourselves in that place and imagine what it would be like. We knew that there were people in their beds. We knew that they needed to be attended to at night. . . . So it was, basically, just a simple logical next step. We thought . . . there had to be hundreds or thousands of doctors, working at night, actually risking their lives. They need light. There were emergency relief workers, genuine heroes, risking their lives to help save others. And how can you do this in utter darkness? You just can't. That's how the slogan came about. (pers. comm. 2018)

This slogan—"You can't fight Ebola in the dark"—offered a powerful, urgent rationale for solar powered lighting in the middle of the epidemic. But it also established a causal relationship between the actions of European solar entrepreneurs and the outcomes for people they would never meet in West Africa, between the manufacturing and distribution of their brand name solar powered products and human life, and it cast the solar lamp, the solar company and its founders as heroes in a story of technological salvation.

On the ground in West Africa, the needs for lighting amidst the epidemic were more complex. If biomedical treatment was curtailed at night it was not always, or only, for lack of electric lighting, and there were clear differences between urban and rural treatment centres. Early on in the outbreak, for example, Sierra Leone"s flagship Ebola isolation unit, located inside the capital Freetown's largest government hospital, was actually locked at night because there were no night staff, and doctors could not risk the prospect of patients leaving and infecting others (Street 2019).

Alongside the slogan, the company's online crowdfunding campaign was accompanied by what advertising agencies and marketing professionals call a "hero image." With restricted international travel, the team were unable

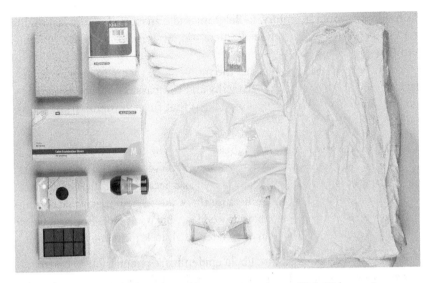

Figure 1. "You Can't Fight Ebola in the Dark", campaign image (WakaWaka 2014).

to source professional images of their solar lighting products being used in context. Instead, they had to create an alternative image; one capable of suggesting their product was part of existing emergency aid kit or was already in circulation and use.

They acquired a pair of yellow rubber gloves, a yellow raincoat, a box of fasteners, a box of latex examination gloves, a sponge, a pair of googles, and a bottle of antibacterial solution with a 70 percent alcohol base from high street hardware stores in Amsterdam. A photographer laid out the items in a lightbox alongside the distinctive yellow case of their lighting products and took a carefully composed shot. The image sought to capture and appropriate the symbolism of epidemiological crisis and emergency for the solar lighting industry before the device had ever been used or deployed in this context. As one of the company's founders put it, "We used that image because we didn't have actual footage coming in of actual doctors in suits helping people, buckets with all kinds of sanitising material and our light being distributed" (pers. comm.).

As transport links between Europe and West Africa began to be cut, the company was contacted by a Norwegian NGO and asked to supply small solar lights for an emergency medical kit that was going to be distributed to medical professionals and volunteers travelling to Liberia and Sierra Leone. The company arranged to have these shipped from the Netherlands, where it stored its product inventory, to Belgium and transported on the last flight from Brussels to Liberia, before all international commercial flights into the country were cancelled.

One of the founders recalled the moment for me in an interview. His narration describes feelings of empathy and compassion, as well as individual heroism and recklessness; an account that fuses humanitarian sentiment with a spirit of entrepreneurship:

> People in western Africa were already plagued by so many troubles. . . . It's a very hard life for people there and especially those people also having to fight this disease. I was so moved by that and also because it was the last instance that we could ever help there, because afterwards the countries would be completely sealed off from the outside world. . . . But it was quite irresponsible of me because we depleted our own commercial stock. (pers. comm.)

The company's response to the Ebola epidemic was impossible to disentangle from their business. The founders hoped that they would be able to convert the visibility of their product during the Ebola crisis into post-Ebola markets for their products. In Liberia, they had meetings with politicians, distributors and non-governmental organisations. "I even had dinner with the president at some point," one of them explained. "We were talking with the first lady and the foundation that she set up."

But the company found it very difficult to press their advantage, lacking personal connections and a physical presence in Liberia's capital Monrovia. To the disappointment of one of the founders, efforts to convert partnerships during the emergency into a more sustainable, long term business in the country failed. "I very much had hoped that I would be able to do that," they told me, "but you have to be very much present on the ground in Africa, you need personal contacts or an actual office with people that are actively on a daily basis there, otherwise it doesn't work."

They remained dependent on non-governmental partners and did not have the resources to establish a local sales team. "We weren't able to do that. . . . There was no way for people to know where to buy more of our products. . . . We didn't have distribution." Rebuffed, the company focused on strengthening and expanding sales in East Africa, where their everyday work demanded attention to production, financing, logistics, social media, and customer support. A year after its efforts to grow a market for its products in West Africa, the company recorded a loss of more than EUR 600,000[1] and, after repeated attempts to raise investment, it declared bankruptcy. In 2018 its assets—including outstanding stock, designs, and its brand name—were bought up by a Dutch business school.

Such attempts to expand markets across sub-Saharan Africa in contexts of humanitarian crisis might appear predatory. But clearly this is not how entrepreneurs, executives, sales managers, market researchers and policy makers make sense of their own actions, or how they connect solar energy to improvements in the lives of unrelated strangers. How then do solar entrepreneurs reconcile or accommodate themselves to the melding of commercial self-interest and care?

Solar Humanitarianism

The promise that our energetic relationships to the sun, mediated by technology, might materialise new relationships to one another has long animated humanitarian action. Indeed, humanitarianism itself is a distinctly carbon form of care, forged at the beginning of carbon fuelled European modernity. The energetic biopolitics—or "energopolitics" (Boyer 2019)—of humanitarianism are rooted in the history of capitalism. What we call humanitarianism today has its origins in the conjuncture of eighteenth-century movements to abolish slavery with new forms of market discipline (Haskell 1985a, 1985b) and in the increasing demand for and combustion of coal as a fossil fuel (Malm 2016). As Cara New Dagget (2019) has shown, European colonialism yoked the science of energy systems and thermodynamics to religious proselytising and a moral, socio-technical project to put colonised bodies and persons to work. For abolitionists, fossil fuelled machines provided both evidence of European cultural and technical superiority, as well as a metaphor for the efficiency and productivity that liberated bodies might achieve as economic subjects (Daggett 2019: 150).

Solar powered humanitarianism involves practices, commitments, and relations that are continuous with this form of care. Just as fossil fuels once connected registers of science and religion in a justification for imperial expansion (Daggett 2019: 151), the promise of solar photovoltaics reconnects them in a rationale for market-led development across Africa. Following World War II, humanitarian sentiment was allied to projects of international development and the expansion of markets for technological innovations in Europe's former colonies (Escobar 1995). With the invention of the modern silicon solar cell, white engineers and solar evangelists from the Global North presented solar photovoltaics as a novel harbinger of radical change in the unelectrified world (Perlin 1999). These early attempts to identify markets for solar photovoltaic systems across "underdeveloped" parts of Africa fused Christian theology with libertarian politics. Off-grid solar systems promised to grant people a direct relationship to the sun, bringing autonomy and freedom from fossil fuels and energy companies, from government and the grid-based infrastructures of modernity, and from the perceived constraints of tradition and culture.

At the turn of the twenty-first century international commitments to achieving universal access to electricity and entrepreneurial investments in off-grid solar power saw humanitarian sentiment embedded in efforts to create new markets at the bottom of the economic pyramid (Cross 2013, 2019). These north-south flows of solar power have attracted increased scrutiny. To some critics, the humanitarian promise of off grid solar energy products in these geographies is an alibi for the expansion of consumer debt, and the reformatting of colonial era economic relationships. But if we are to take seriously people's investment in ethical practice and their concern to do good (Robbins 2013) we might also ask, how is it that the relations that characterize fossil fuelled capitalism are normalized and naturalized by those who also say they want to disrupt and transform our socioeconomic system through the energy of the sun?

Imagine you are about to graduate from a graduate program in business and management, design, or social development. Now, imagine that some of your classmates invite you join them in launching a start-up company that will manufacture and sell ultra-affordable solar powered lighting and charging systems to people living in energy poverty across parts of sub-Saharan Africa. You are faced with a choice: Do you join the solar start-up or not? If you accept, you become a social entrepreneur and there is a possibility that your actions can alleviate the suffering and hardship of others. If you decline—perhaps because you have been socialised to be risk averse or because you are saddled with considerable student debt—are you also choosing not to act? By not acting are you condemning people to a continued life of suffering in energy poverty?

This is the precisely the kind of ethical dilemma posed by the economic historian Thomas L. Haskell in a classic, two-part essay on capitalism and humanitarian sensibilities (1985a, 1985b). In it, Haskell developed a critique of humanitarian action that sought not to unmask the compassion of others as unexamined self-interest or social control. Instead, he invited readers to reconsider the historic relationship between economic and moral logics in capitalist societies without "reducing humane values and acts" to epiphenomena (1985a: 341).

The expansion of market discipline in Europe between the seventeenth and eighteenth centuries, Haskell argued, brought shifts in "conventions of moral responsibility"—a "new constellation of attitudes and activities"—that underpin what we have come to call humanitarianism (1985a: 342). At the heart of these shifts, Haskell suggested, were new "perceptions of casual relations" between the actions of an individual and the effects on an unrelated or distant stranger. The gradual elaboration of techniques for the conduct of everyday business—from legal contracts that guaranteed the extension of trust to strangers" (1985b: 556) to methods for long range forecasting or planning (1985b: 558)—Haskell contended, established new ways of connecting actions and consequences over time and space. One outcome, he wrote, was that "teaching people the virtues of reflection and close attention to the distant consequences of their actions came to be regarded as a universal key to social progress" (1985b: 561–62).

Haskell's argument has become a key reference point in studies of humanitarianism across fields of anthropology, history, development, and cultural studies (Minn 2007; Redfield 2013; Bornstein 2012). If the relationship between private actions and humanitarian goods can sometimes appear to be a recent phenomenon, Haskell's work is used to remind us that they have a longer history. As Tom Scott-Smith (2016) puts it: "Humanitarians have long sought solutions in the private sector, been attracted to new technologies and purchased goods through the market, and the relationship between entrepreneurship, capitalism and philanthropic sentiment goes back to the very origins of modern humanitarianism."

Haskell's analysis goes further, however, and has more to offer an understanding of humanitarianism, technology, and capitalism. As he showed, humanitarian sentiments are not static, they can be transformed by new ideas, technologies, and institutions that alter the capacity of people to act at a distance.

Haskell outlined four preconditions for the emergence of humanitarian sentiments. Before people can feel obliged to aid distant strangers, he proposed, they must first agree on an ethical maxim that makes helping distant strangers the right thing to do. Second, people must perceive themselves to be causally involved in an evil event. Third, people must be able to perceive a causal connection, or a chain of cause-and-effects, that begin with their actions and end with the alleviation of the stranger's suffering. Finally, and perhaps most importantly, Haskell proposed, "the recipes for intervention" that are available to people "must be of sufficient ordinariness, familiarity, certainty of effect, and ease of operation that our failure to use them would constitute a suspension of routine, an out-of-the-ordinary event, possibly even an intentional act in itself" (1985a: 358).

Over the past twenty years, the conditions have been laid for the emergence of solar humanitarianism, in which the work of manufacturing and distributing solar powered products are linked directlty to the alleviation of suffering for people living in energy poverty. The UN's Millennium Development Goals and the Sustainable Development goals have provided an ethical rationale for acting on the energy poverty of people living without electricity. A deepening awareness of anthropogenic climate change, coupled with a greater understanding of the specific role of European industrialisation in accelerating global heating, has given rise to new feelings of white guilt and complicity (Markowitz and Shariff 2012). Meanwhile, new rationales or justifications for the positive effects of private enterprise on others, and a new spirit of entrepreneurship has transformed the ways in which some people feel compelled to intervene in the lives of others.

Against this backdrop, we can discern two distinct recipes for intervening in the world, for acting as a solar humanitarian. The first involves acts of giving, likely mediated by digital technologies and online systems. As Anke Schwittay (2019) describes, drawing on Haskell's work, the act of clicking through a series of website links or entering payment data online to make a donation to a charitable organisation, or to lend money to people in need, has become an ordinary, familial recipe for "everyday humanitarianism." Such recipes for intervention have become a vital part of solar humanitarianism. They underpinned WakaWaka's campaign, as it invited people to support their work in Liberia and Sierra Leone during the Ebola epidemic by pledging a minimum US$9 via the Kickstarter crowdfunding platform. Six years later, in the midst of the global Coronavirus pandemic, they underpinned a call to action by UK-based solar energy entrepreneur Jeremy Leggett for people to support his charity, Solar Aid, in rural Zambia.

"Please would you join me today in funding a box of solar lights," he wrote to the charity's supporters in April 2020, "with your family name written invisibly on it, for a health clinic in sub-Saharan Africa? I, my compadres in SolarAid, and many Africans would be most enormously grateful. Our fundraising campaign page tells you how you can do that with a few clicks" (Legget 2020).

The second involves acts of entrepreneurship. The work of entrepreneurship—developing a business plan, pitching to investors, securing finance, setting up logistical systems for mass manufacturing and distribution, managing risk and people—has become a preferred mode of late capitalist humanitarianism. Indeed, for some entrepreneurs doing humanitarianism has been internalised as the ethically imperitive mode of doing business. Today, entrepreneurship itself is a recipe for humanitarian action: as ordinary and familial a recipe for intervening in the world as an act of charity; so common-sensical that the failure to use it would be a cause of continued suffering. As Imre Szeman (2015: 472) has put it, entrepreneurship is "the new common sense":

> Entrepreneurship has come to permeate our social imaginaries in a way that has quickly transformed its claims and demands on us from fantasy into reality. We are all entrepreneurs now, or, at a minimum, we all live in a world in which the unquestioned social value and legitimacy of entrepreneurship shapes public policy, social development, economic futures, and cultural beliefs and expectations.

Over the past twenty years this spirit of entrepreneurship has found a particular outlet in the green economy, underwriting green capitalism (Mazzucato 2015). As Jesse Goldstein (2018: 3) puts it, emotional and financial investments in the promise of technological salvation through renewable energy, like solar, have shaped, the "contours and trajectory of green technological innovation."

This juxtaposition of solar energy and entrepreneurship has provided new ways of acting at a distance, of coming to the aid of distant strangers. In a context in which it is possible to help design, manufacturer, or distribute solar powered technologies, in which entrepreneurship and the expansion of market relations appear as the solution to the problem of energy poverty, the failure to become a solar entrepreneur, if you have the capacity to do so, becomes ethically almost indefensible.

Solarity after COVID-19

Against the backdrop of COVID-19, there seems much to learn about solarity in the locally organised or community-based networks of mutual aid that have emerged to ensure that people in self-isolation can continue to access sunlight (either directly or mediated through fruit, vegetables, and grains). In the midst of a global shutdown, in which the carbon-grid world is suddenly disabled, these forms of solarity appear to offer us some sense of what post-carbon futures might look like. But amidst the pandemic perhaps there is also something to learn about solarity in the continued expressions of care for distant others: people far away, unconnected by place or kinship by those who also seek to maintain and expand market relations.

The social and economic impact of COVID-19 has led to disruption in global supply chains for solar photovoltaic technologies, from delays in the supply of multi-crystalline solar wafers and components to increases in the prices of key components and manufacturing timelines. Across sub-Saharan Africa, lockdown measures and economic uncertainty are creating a collapse in consumer demand for off-grid solar energy products.

In March 2020, two weeks after the World Health Organisation declared a global pandemic, I joined an online webinar hosted by a global solar energy industry body, with representatives of solar manufacturing and distribution companies. How can governments and investors best ensure that people living in chronic poverty continue to have access to solar powered lighting and charging, they asked? Should the industry ask governments to provide subsidies to off grid solar companies directly, to protect jobs and allow them to continue manufacturing? Or should governments be providing subsidies direct to consumers, to allow them to continue buying products?

One discussant, the representative of a publicly funded European development agency, ventured their opinion. "It is a tricky subject, of course . . . because we are still aiming to build markets, long term sustainable markets. We all hope this is a temporary crisis, but we still want to build markets."

If such commentators appear to turn a blind eye to the impact of these interventions—appearing to ignore, for example, how these interventions sustain or reinforce the unequal distribution of wealth, creating new dependencies, or fostering inequities—perhaps it is less because they are engaged in deliberate acts of deception (or even forms of self-deception) and rather, as Thomas Haskell argued, because this is the limit of their humanitarian capacity, a reflection of their ability to focus on one quantum of suffering at a time.

The solar industry's specific attention to quantifiable suffering— the deprivation, privation and hardship endured by those living without electricity—and its focus on actions that extend market relations has served to establish a set of conventions for moral work and responsibility; allowing entrepreneurs to direct their humane acts at a specific, narrowly defined problem and affording them a modest, perhaps even temporary kind of ethical shelter from other concerns.

For some critics, "carbon forms" (Yarina 2019)—that is, the legacy of carbon-fuelled modernity as it is manifested in the built environment, in technology or, we might add, in humanitarian sentiments—can have no place in the revolutionary adaptations required by climate change. But if we want to understand how the relations that characterised fossil fuelled capitalism can be transformed then we need to recognise how carbon forms of care can be normalized and naturalized, made all light and clean, by sunshine.

References

Balch, Oliver. 2014. "Five Products that Help the World's Poorest People." *Guardian*, November 10. theguardian.com/sustainable-business/2014/nov/10/consumers-shopping-solar-ebola-gucci-water-giving-tuesday.

Bornstein, Erica. 2012. *Disquieting Gifts: Humanitarianism in New Delhi*. Stanford: Stanford University Press.

Cross, Jamie. 2013. "The 100th Object: Solar Lighting Technology and Humanitarian Goods." *Journal of Material Culture* 18, no. 4: 367–87.

Cross, Jamie. 2019. "The Solar Good: Energy Ethics in Poor Markets." *Journal of the Royal Anthropological Institute* 25, no. S1: 47–66.

Escobar, Arturo., 1995. *Encountering Development: The Making and Unmaking of the Third World*. Princeton: Princeton University Press.

Goldenberg, Suzanne. 2015. "Coal Giant Exploited Ebola Crisis for Corporate Gain, Say Health Experts." *Guardian*, May 20. theguardian.com/environment/2015/may/19/peabody-energy-exploited-ebola-crisis-for-corporate-gain-say-health-experts.

Goldstein, Jesse. 2018. *Planetary Improvement: Cleantech Entrepreneurship and the Contradictions of Green Capitalism*. Cambridge: MIT.

Gunn, Wendy, Ton Otto, and Rachel Charlotte Smith, eds. 2013. *Design Anthropology: Theory and Practice*. London: Bloomsbury.

Haraway, Donna. 1987. "A Manifesto for Cyborgs: Science, Technology, and Socialist Feminism in the 1980s." *Australian Feminist Studies* 2, no. 4: 1–42.

Haskell, Thomas L., 1985a. "Capitalism and the Origins of the Humanitarian Sensibility, Part 1." *American Historical Review* 90, no. 2: 339–61.

Haskell, Thomas L., 1985b. "Capitalism and the Origins of the Humanitarian Sensibility, Part 2." *American Historical Review* 90, no. 3: 547–66.

Leggett, Jeremy. 2020. "Imagine Facing This Pandemic in the Dark." *Future Today* (blog), April 10. jeremyleggett.net/2020/04/10/imagine-facing-this-pandemic-in-the-dark-thats-what-much-of-rural-zambia-and-malawi-will-have-to-do-and-we-can-do-something-about-it/.

Malm, Andreas. 2016. *Fossil Capital: The Rise of Steam Power and the Roots of Global Warming*. London: Verso.

Markowitz, Ezra M., and Azim F. Shariff. 2012. "Climate Change and Moral Judgement." *Nature Climate Change* 2, no. 4: 243–47.

Mazzucato, Mariana. 2015. "The Green Entrepreneurial State." In *The Politics of Green Transformations*, edited by I. Scoones, M. Leach, and P. Newell, 152–70. London: Routledge.

Minn, Pierre. 2007. Toward an Anthropology of Humanitarianism. *Journal of Humanitarian Assistance*, August 6. sites.tufts.edu/jha/archives/51.

Perlin, John. 1999. *From Space to Earth: The Story of Solar Electricity*. Ann Arbor: Aatec.

Redfield, Peter. 2013. *Life in Crisis: The Ethical Journey of Doctors without Borders*. Berkeley: University of California Press.

Robbins, Joel. 2013. "Beyond the Suffering Subject: Toward an Anthropology of the Good." *Journal of the Royal Anthropological Institute* 19, no. 3: 447–62.

Rowe, Peter G. 1987. *Design Thinking*. Cambridge: MIT.

Schwittay, Anke. 2019. "Digital Mediations of Everyday Humanitarianism: The Case of Kiva.org." *Third World Quarterly* 40, no. 10: 1921–38.

Scott-Smith, Tom. 2016. "Humanitarian Neophilia: The 'Innovation Turn' and Its Implications. *Third World Quarterly* 37, no. 12: 2229–51.

Street, Alice. 2019. "The Limits of Medical Heroism: Reflections on Getting to Zero." *Somatasphere*, May 21. somatosphere.net/2019/the-limits-of-medical-heroism-reflections-on-getting-to-zero.html/.

Svec, Vic. 2015. "Peabody Energy's Discussion of Africa's Ebola Crisis Was Perfectly Proper." *Guardian* (Letters), May 25. theguardian.com/environment/2015/may/25/peabody-energy-discussion-of-africa-ebola-crisis-was-perfectly-proper.

Szeman, Imre. 2015. "Entrepreneurship as the New Common Sense." *South Atlantic Quarterly* 114, no. 3: 471–90.

World Bank. 2020. "The 2020 Global Off-Grid Solar Market Trends Report." *Lighting Global*, February 18. lightingglobal.org/resource/2020markettrendsreport/,

Yarina, Lizzie. 2019. "Towards Climate Form." *Log* 47: 85–91.

Sheena Wilson

Solarities or Solarculture:
Bright or Bleak Energy Futures
and the E. L. Smith Solar Farm

This article explores the possibilities of the verb
to solarize through a case study of the proposed
E. L. Smith Solar Farm in Edmonton, Alberta. I
study this solar farm proposal in local, provincial, and federal contexts, often at odds. On the one
hand, oil-loyal[1] discourses foment negative political
and media responses to people and enterprises
deemed anti-oil; this was particularly the case with
the infrastructure blockades by those demanding
withdrawal of the Teck oil sands proposal of late
2019 and early 2020. On the other hand, Indigenous decolonization and reconciliation efforts
intensify as people at the intersections of gender,
race, class, and ability continue to demand equity
and a voice at the table when it comes to proposed
energy infrastructure and politics—whether the
site of tension is another proposed oil sands mine,
a pipeline to get (more) Canadian crude to tidewa-
ter, or a solar installation.

The E. L. Smith Solar Farm provides an
interesting case study of the material and social
realities undergirding solar fantasies, in part
because it illustrates the distinction between solarcultures and solarities.[2] Each of these concepts
is founded on material infrastructures of solar
energy. Whereas solarcultures tend to reproduce

The South Atlantic Quarterly 120:1, January 2021
DOI 10.1215/00382876-8795779 © 2021 Duke University Press

the structures of inequality that characterize the petrocultural regimes they otherwise purport to replace, solarities, by contrast, comprise an intersectional, equitable, and ethical response to those regimes. To solarize is to contest and subvert, rather than to reproduce, the material relations of petroculture. The risk (from a social justice perspective) is that solarculture can easily stand in for solarity, especially under environmental conditions in which rapid transition to renewable energy sources is a real imperative.

Land(ing) Solar Projects

Following the public hearings for the E. L. Smith Solar Farm presentation to Edmonton's city council in 2019, I conducted a set of research interviews as part of the *Deep Solarities* podcast series. The E. L. Smith Solar Farm is a solar energy infrastructure project proposed for installation at the E. L. Smith Water Treatment Plant site located on the banks of the North Saskatchewan River in Edmonton's River Valley, the largest urban greenspace in North America "at 22 times the size of New York City's Central Park" (Explore Edmonton n.d.). The project involves a ten-megawatt solar installation that, if approved, will contribute to the City of Edmonton's overall goal of producing 10 percent of its electricity locally from renewables and offer the daily benefit of lowering the plant's carbon footprint (EPCOR 2018).

The project initially appeared to check all the relevant boxes for a renewable energy project. In late 2018, as the proposal was being prepared for city council's consideration, it received an injection of funding: C$10.7 million from a federal program for green infrastructure projects and another C$1.9 million from Alberta Innovates. There had been significant efforts made to ensure that the project had the necessary social license. For example, the plan included an Indigenous herb garden, educational opportunities for postsecondary students, and public access initiatives (EPCOR 2018). Enoch Cree Nation, whose reserve lands border the municipality, had also signed on in support of the project. However, by the time the E. L. Smith proposal went to a public hearing before city council on June 17, 2019, the project had become increasingly contested by conservation groups, Enoch had withdrawn its support, and the proposal was referred back to city administration and EPCOR, the city's utility company.[3]

In Canada, in Alberta, and certainly in Edmonton, the battle lines for the future are being drawn around energy issues, with some rallying to extend oil extraction and build more pipelines, while others stand on the front lines of the climate justice movement demanding action that responds not only to the Paris Agreement (2015) and the IPCC special report (2018), but also to the ninety-four recommendations made by the decade-long Truth and Reconciliation Commission (2015). Edmonton is the provincial capital located just south of the Alberta oil sands, the world's largest bitumen deposits. Despite being situated in the seemingly oil-loyal province of Alberta, Edmonton has relatively progressive energy and climate politics. Its nonpartisan city council has been increasingly taking action on climate for the past decade. In 2015, council approved the Community Energy Transition Strategy, a commitment to action on global climate targets. The strategy guides policy decisions and is mobilized to shift public attitudes towards climate and energy transition via the city's Change for Climate campaign.[4] In 2018, Edmonton hosted the IPCC Cities and Climate Change Science Conference, where the mayors of close to thirty-four hundred North American municipalities signed on to the Edmonton Declaration, formalizing their commitments to take action toward achieving the Paris Agreement (City of Edmonton 2018). All of this to say that despite mainstream media news coverage of Canadian politics internationally, and Alberta provincial politics nationwide that paint Albertans as oil-loyal and against energy transition, many Canadians, Albertans, and certainly Edmontonians (74 percent according to the latest municipal polls) are concerned about climate change and want to take action (City of Edmonton 2019).

When the E. L. Smith Solar Farm proposal went to city council in June 2019, action on climate change was on both the formal agenda and Edmontonians' minds. As the proposal was making its way through the stages of approval in 2018–19, so was the *Climate Resilient Edmonton: Adaptation Strategy and Action Plan* (City of Edmonton 2015a), which outlines coming climate challenges to the municipality, making very clear the need to act urgently. Over the summer, city council debated the need to increase their commitments to addressing climate change by revising the Community Energy Transition Strategy (Johnston 2019). These debates on how, and how fast, to act on climate change culminated two months later on August 27, 2019, when Edmonton City Council voted ten to two to declare a climate emergency (Metz 2019). In so doing, Edmonton joined "over 1390 local governments in 27 countries [that] have declared a climate emergency and committed to action to drive down emissions at emergency speed" (The Climate Mobilization n.d.). The motion was put forward by the city's only Indigenous city councillor, Aaron Paquette; however, it is fair to say that it would not have been successful without the leadership and collaboration of local climate activist groups and the participation of diverse community members—"young people to grandparents"—who showed up in large numbers to provide support (Issawi 2019). It is in this local context that the proposal for the E. L. Smith Solar Farm project was moving forward.

Mobilizing toward Solarities by Mobilizing against E. L. Smith Solar Farm

Despite initial enthusiasm, over the course of 2018–19, resistance to the E. L. Smith Solar Farm project began to build. Resistance came from long-time solar advocates and environmental organizations—such as the Sierra Club Canada, the North Saskatchewan River Valley Conservation Society, and Edmonton River Valley Conservation Coalition—who did not want to see the land rezoned from Metropolitan Recreational to Direct Development Control Provision (a zoning category used for industrial development related to providing utility services) out of concern for setting a precedent that would lead to further industrial development in the River Valley (see also Richmond 2020). Their resistance to the solar project came from their commitment not only to protect the area for human recreational purposes, but also to ensure multispecies flourishing and fight against proposed disruptions to the ecosystems and the elimination of wildlife corridors. As one interviewee put it, many of these stakeholders have been advocates for energy transition, but not at the expense of biodiversity (Feroe 2020). Implicitly, despite sometimes divergent positions, these environmental organizations largely lean toward visions of solarities, rather than solarcultural futures. This municipality, somewhat anomalous in provincial and federal contexts, has strong local political networks grounded in community leagues with commitments to labor and environment as well as specific local issues, one of which has always been protecting the River Valley that links communities within and beyond the municipality. Citizens, working at the community level and with a range of environmental organizations including the aforementioned, have stood guard for decades to protect the River Valley, specifically against the threat of industrial encroachment into the area.

As different stakeholders debated land use—and, ultimately, whose rights take precedence—a third-party contractor was hired to complete an environmental impact assessment as part of the rezoning process. This investigation built on an earlier report that identified Indigenous archaeological remains in the area, which suggested possible designation as an important heritage site (Alberta Culture and Tourism 2018; see also Sharphead 2020). Canada is a country made possible by colonization and the dispossession of lands, and this new solar farm is being proposed in lands covered by Treaty No. 6, an agreement signed in 1876 that covers an area of 121,000 square miles in the central regions of what are now the provinces of Alberta and Saskatchewan. Indigenous signatories and their descendants understand the treaty to be an agreement to share the land and resources, although the Crown operates largely as though this was a surrender agreement. *All energy projects, green or fossil, happen on, to, and with the land.* The area in question, although now part of the municipality, was historically part of the reserve lands of the Enoch Cree Nation, which borders the city; this First Nation was dispossessed of these lands through yet other "dubious surrender processes" in 1908 (Houle 2016). By the time of the June 2019 hearing, Enoch Cree Nation had withdrawn their approval for the project. At this hearing, several stakeholders spoke out about the issues outlined above, but most significantly, when the issue of Indigenous archaeological findings was raised, the public hearing was postponed by city council. In response, Jason Kenney, then the newly elected premier of Alberta, seemingly decided to make policy on Twitter. On June 19, he posted, "I agree with these Edmonton residents & the Enoch 1st Nation. The River Valley should be a ribbon of green, free of industrial projects. That's why our govt is ending funding from this solar farm in the Valley & will help create the nearby Big Island Park" (Kenney 2019). With this tweet, Kenney aimed to undermine the development of the solar project while seeming to perform allyship. What complicates public understanding of Kenney's United Conservative Party's (UCP) hard-line energy politics are instances such as this, where a vision for solarities (equitable, decolonial, bidoverse decarbonized futures) plays into his ambition to reinvigorate Alberta's petrocultural state. Political spin allowed for his ongoing refusal of energy transition to appear as support for environmental and Indigenous issues.

Performances of Petrocultural Allyship vs. Solarities

In addition to the context given so far, it is worth noting that the E. L. Smith Solar Farm was being prepared for presentation to city council within a global context of mass mobilizations demanding climate action in solidarity with both Greta Thunberg and the youth climate movements, as well as blockades going up across Canada in 2020 in support of the Wet'suwet'en Nation and the hereditary chiefs' refusal to let the Coastal GasLink pipeline pass through their unceded territories in northern British Columbia. On the heels of having supported Enoch Cree First Nation against the E. L. Smith Solar Farm, the Alberta UCP government responded to the Wet'suwet'en by reducing the conversation to lost revenues and jobs to set the stage for Bill 1 (Government of Alberta 2020).

As the Legislative Assembly of Alberta opened session on February 25, 2020, the first bill Kenney tabled was Bill 1, the Critical Infrastructure Defence Act. It went to second reading on February 26, and "imposes stiff new penalties on law breakers who purposefully block critical, essential infrastructure, such as railways, roadways, telecommunication lines, utilities, oil and gas production and refinery sites, pipelines, and related infrastructure" (Kenney 2020). Beyond the infrastructure Kenney named in media reports, the legislation defines essential infrastructure very broadly to include almost anything, and penalties range from "$10,000 for a first offence, $25,000 for a subsequent offence, and up to six months in jail" (Kenney 2020; see also Kelly and Nassar 2020). In the context of climate justice initiatives and protests in Canada and across North America, this has to be understood as a law that specifically targets Indigenous Peoples and environmental activists who are challenging business as usual by refusing new energy projects that respond to neither reconciliation nor climate change. More broadly, this law threatens the political agency of all Albertans or anyone exercising their democratic right to protest in Alberta.[5] Kenney and all the members of the Legislative Assembly who voted with him, including some of the opposition, are making it dangerous for citizens to stand in support of the Wet'suwet'en, one another, and their political leaders demanding climate action, like Edmontonians did to support Councilor Paquette's motion to declare a climate emergency. As citizens refuse oil-loyal politics and organize to demand change, the Alberta government is passing legislation to outlaw our capacity to stand together—to create solidarities and solarities.

Working in concert with the leader of the federal Conservative Party of Canada, Andrew Scheer, and his federal party counterparts, Kenney shaped public opinion by speaking to the media and the public about the illegality of blockades by Indigenous Peoples and supporters of Indigenous nation-to-nation relations. He followed up by then making it so: by writing it into law at the provincial level in Alberta and passing the Bill 1 legislation. This move, among others, by Kenney and his UCP, foments racist politics and reinforces the myth that oil (the rhetorical stand-in for all fossil energy sources) is a founding characteristic of the nation and the key to its future (see Jacobs et al. 2020).

In fact, *what is foundational* to the legal legitimation of the nation are the international treaties signed between the Crown and the many Indigenous nations, not to mention the illegitimate takeover of lands never ceded by

Indigenous nations. Oil-loyal political discourse constructs Indigenous Peoples, Indigenous land claims, and Indigenous political organizing as obstacles to resource extraction and thereby traitors to the nation-state, when in fact the nation only exists because of the peace and friendship treaties signed with these nations, which have now been run roughshod by representatives of the Crown and Canadian governments for over three hundred years.

In short, allyship for Kenney is a performance that he engages in only when allegiances serve his larger and consistent purpose: to defend and sustain petrocultural interests, beyond all reason, even when oil is trading at $-37.63 a barrel and dropping (Chapa 2020). Kenney's relations with Indigenous land and conservation politics are entirely organized around exploitation; his commitment is to multinational fossil energy interests and ensuring their continued ability to extract profit from land. The other necessary exploitation, however, is that of Indigenous Peoples and their politics. Kenney aligns where it serves his purposes and he writes legislation to overrule Indigenous protest and non-Indigenous solidarity actions when it does not. Non-Indigenous peoples are also consistently exploited in the interest of aligning the public will with extracting profits from land, which is necessarily entangled with the exploitation of others: human and more-than-human. Likewise, the citizenry is being made increasingly reliant on a fossil-energy industry that is fast becoming globally obsolete as green tech improves and the urgency to respond to climate change escalates with pressure from across the political spectrum, especially youth movements. A democratic order undergirded by petrocapital renders all citizens precarious—the very opposite of what a future informed by solarities could mobilize.

Control of not only land and resources but also bodies has been achieved through ongoing colonization, or more specifically, petrocapitalism that is also, fundamentally and inescapably, patriarchal, ableist, and cisheteronormative. This worldview produces the pillage of resources as well as the oppression of whole swaths of the global population; in Canada, it underlies the sociological phenomenon of missing and murdered Indigenous women in Canada.[6] It is also expressed daily and continuously as misogyny and racism directed at undermining the disproportionate numbers of women, members of the LGBTQ2SIA+ communities, people of color, and Indigenous Peoples on the frontlines of the climate crisis, demanding justice. The misogynist dismissal of Greta Thunberg is one example (Wilson 2020), and the media's representation of Indigenous activist Chief Theresa Spence is another (see Wilson 2017). As the age of fossil-fuel exploitation ends, Albertans are left with the abandoned wells, the turned-over boreal forest, and the ruins of an age gone by, while much of the wealth has left the province, following the networks of petrocapitalist infrastructure around the world.

To achieve solarities (bright solarities), rather than solarcultures (bleak solarities),[7] new energy sources and systems need to be implemented in ways that are adequate to the situation at hand. Climate change is a symptom, not the cause, of our current planetary crisis. This means thinking and acting meaningfully as every new energy project is put *onto the land*. Whether, like at E. L. Smith, the project is solar or a new fossil infrastructure, our responses to climate change must be about more than decarbonizing the atmosphere. The real challenge is to address the bankrupt worldview and economic imperatives that push the planet to 2.5 times its limits. This starts by thinking big and thinking differently—thinking solarities—even with small local projects like E. L. Smith, because one thing COVID-19 has taught us, if globalization hadn't already, is that the planet is an interconnected ecosystem and what we do locally has ramifications beyond even what we seem capable of understanding or imagining. We need to transform our imaginaries of who we are in relationship to this planet and to what futures are possible. And we must avoid retreating to small-minded thinking grounded in scarcity mentalities that foment the violences of the current petrocultural age.

With politics dominated the world over by increasing numbers of governments flouting democracy and maintaining power through the silencing and dispossession of voices and bodies who dare to resist political economies, the need to solarize is pressing. Throughout February 2020, the Canadian media reported continuously on blockades, police brutality, and attacks on Indigenous Peoples, climate protesters, and those putting their bodies and lives on the front lines of the climate justice movement. On February 23, 2020, Teck Resources Limited withdrew its proposal for a massive project in the Northern Alberta oil sands. In a letter to Minister of Environment and Climate Change Jonathan Wilkinson announcing the withdrawal of the Frontier proposal, Teck CEO and president Don Lindsay wrote that Teck is committed to producing energy in ways that respond to climate change, and that "the promise of Canada's potential will not be realized until governments can reach agreement around how climate policy considerations will be addressed in the context of future responsible energy sector development" (Lindsay 2020). In short, because the cost of carbon pricing has not yet been set, it makes for uncertain profit margins. Furthermore, when the externalities of fossil energy are no longer subsidized by oil-loyal governments, and when the detritus of industrial production is no longer borne by the human and more-than-human communities and ecologies it disrupts free of charge, these companies will no longer extract the excessive profits of a foregone era, securing power, writ-large and understood as both energy/energy systems and as social relation.

At the announcement of the project being cancelled, Alberta premier Kenney deflected blame away from his provincial government's unwillingness to respond to federal political imperatives by setting a price on carbon, instead turning attention to Trudeau, claiming the project failed because of the federal government's inability to control protests across the country: "It is what happens when governments lack the courage to defend the interests of Canadians in the face of a militant minority," said Kenney (Rieger 2020). However, his reference to a "militant minority" is increasingly not the case. Millions of people the world over are demanding change, and here in Canada, there is significant support for the Wet'suwet'en cause.[8] However, what he is really demanding is that Canadians refuse to respect the international treaties that are foundational to the creation of this country, in favor of legalizing what until now is really a myth of nation: that we are a people of oil, past, present, and future. In demanding this refusal and outlawing these solidarities, he also refuses just futures, demanding that the petrocultural legacies of injustice will continue to define the future.

Conclusion: Solarities Not Yet Foreclosed

Within this fraught social, economic, and political landscape, how might the E. L. Smith Solar Farm proposal become an example of alternatives to petrocultures? As of spring 2020, the E. L. Smith Solar Farm proposal has yet to come back to city council for a second hearing, and it is unclear how it will be resolved. In the interim, it provides an interesting case study for what it means to solarize, or not. In February 2020, an important stakeholder who had been vocal against the project at the hearing suggested things were "going better" (pers. comm.). What does *better* look like?

Energy justice is not a guaranteed outcome of energy transition. In Canada, Indigenous rights, human and more-than-human rights, resource development, energy transition, and climate action must all be reconciled and dealt with simultaneously. Take, for example, the Teck CEO's announcement of the purchase of a new solar project—the SunMine solar energy facility in Kimberley, British Columbia—that had many in the green energy sector celebrating (Teck 2020). A relatively small project, it seems more like a good PR move than a move from fossil investments to solar. However, it is also a reminder that those concerned about energy and climate justice must agilely follow the battle for the energy future and where it is being fought. Despite Kenney's best efforts to the contrary, it seems, at least recently, that energy futures are shifting away from fossil infrastructure. However, simply shifting the site from which Teck extracts stakeholders' profit margins is not a win for justice. It harkens to a solarcultural future and not anything like a deep commitment to solarities.

Edmonton could set a precedent for how new projects are handled, given that so many municipalities are now taking on a new role in energy leadership. Fossil energy projects are usually located far from urban sites, often in areas that fall under the jurisdiction of provincial and federal politics, which means that municipalities, like their citizens, are quite far removed from the power and politics that they benefit from. Now, as more and more renewable energy sources move closer to or even within city limits, the decision-making roles and responsibilities of municipalities are transformed. Edmonton, in this case, could look to the historical and ongoing practices of oppression by other levels of government and do things differently. It won't just be a matter of whether the E. L. Smith Solar Farm is pushed through on an Indigenous heritage site (as it was suggested might happen), but a matter of thinking through just and equitable options if it is not. Where, then, does the solar farm go? To answer this question demands that those who decide and those of us who give our approvals, whether actively or through complacency, need to think about our relationships to land and to a future defined by solarities.

When we solarize, energy becomes a commons: a shared entity that we draw from as needed and not more. Under solarities, communities will share access to land and resources, and ownership will not endow the right to exploit for personal interest; equitably distributed profits, the benefits (monetary or otherwise) from new projects on the land, will be shared by communities. Energy justice, in cases like E. L. Smith and in all approvals for energy projects, will become the metrics upon which we evaluate viability. In a future informed by solarities, profit will no longer be measured in terms of dollars gained or lost; the parameters of language will shift to terms like *community benefits* and/or *wellbeing economies*, among others, where the primary considerations will be whether everyone in our society has their needs provided for; where wellness will index a highly educated population generating locally specific and internationally applicable knowledge; where the health of communities and ecosystems will be a measure of success; and where health will refer to both physical and social well-being produced through good relations—Indigenous and non-Indigenous, human-to-human, as well as human and more-than-human.

Under solarities, our responsibilities as treaty people to one another and the lands on which we live will override the energy intensive desires that exceed need. In short, to make solarities possible, those with power—energy providers, but importantly policy makers, city councilors, government administrators, utility operators, engineers, architects, city planners, those working in oil and related industries, and the general citizenry demanding more from their leaders—must learn from climate youth activists and Indigenous peoples and communities how to forge solidarities, and how to step up and step aside when it's demanded. Ultimately, the transformations adequate to climate change require a shift away from the modes of being and doing that produced the crisis, which means learning from and with those individuals and communities demanding and defining climate justice. Intersectional knowledge and multidirectional learning will help us to solarize: to mitigate and adapt to climate change in ways that create more equitable and just futures for us all.

Notes

This article was written in early 2020; it went to press before any vote or decision was taken on E. L. Smith by Edmonton's city council.

1   *Oil-loyal* is a term I use to describe an identity politics that aligns oil with what it means to be Canadian and Albertan, and by proxy, insinuates it is treasonous to support climate or energy transition politics. This attitude has been fostered by multiple levels of government as well as corporate media and PR campaigns.

2   Solarities is a neologism: *solar* with *solidarity* to assert that all energy projects, green or not, need to be organized with social justice commitments at the forefront. By expanding multispecies relations to include not only new constellations of kinship but stellar constellations as kin, solarities demand from us humans a more expansive understanding of our situated lives, individually and collectively, and a shift in mindset toward new ways of being and doing that will inform new politics (Mirrorland Collective 2020). Solarcultures, on the other hand, colonize the sun and perpetuate the injustices of petrocultures, albeit fueled by less carbon-intensive energy sources.

3   The details of the motion to send the proposal back for more work is publicly accessible via the City of Edmonton's website: edmonton.ca/residential_neighbourhoods/neighbourhoods/rezoning-proposed-e-l-smith-water-plant-solar-farm.aspx.

4   This campaign and transition strategy can, at least in part, be credited to the work of the Alberta Climate Dialogues (2010–16) research team, led by University of Alberta professor Dr. David Kahane, which ran a series of deliberative democracy consultations on climate change that would give license to the Change for Climate Campaign run by the City of Edmonton, a communications strategy that aims to mobilize *Edmonton's Energy Transition Strategy* and the *Climate and Adaptation Strategy*. See City of Edmonton 2015b, 2017.

5   Alberta's Critical Infrastructure Defence Act extends the ambiguities set by Stephen Harper's federal government in 2015, when it passed Bill C-51, the Anti-Terrorism Act, which rewrote the definition of what constitutes a threat to national security to include a threat to the economy (Theodorakidis 2015).

6   For more on the connection between petrocolonialism and the crisis of Missing and Murdered Indigenous Women (MMIW), see Simpson 2014.

7   The distinctions "bright" and "bleak" that are mobilized here and below are taken from discussions started in the After Oil 2: Solarities school held in May 2019 in Montreal, from which this volume emerges.

8   In mid-February 2020, Ipsos published poll results that said 39 percent of Canadians believe the protests are legitimate and justified, which is a significant percentage (Bricker 2020). Since that poll was conducted, there have been solidarity protests across the country.

References

Bricker, Darrell. 2020. "Majority of Canadians (61%) Disagree with Protestors Shutting Down Roads and Rail Corridors; Half (53%) Want Police to End It." *Ipsos*, February 19. ipsos.com/en-ca/news-polls/Majority-Canadians-Disagree-With-Protestors-Shutting -Down-Roads-And-Rail-Corridors-And-Half-Want-Police-To-End-It.

Chapa, Sergio. 2020. "Oil Could Go to Negative $100 per Barrel in May, Experts Say." *Houston Chronicle*, April 22. houstonchronicle.com/business/energy/article/Expert-Oil-could -go-to-negative-100–per-barrel-15216706.php.

City of Edmonton. 2015a. *Climate Resilient Edmonton: Adaptation Strategy and Action Plan.* edmonton.ca/city_government/documents/Climate_Resilient_Edmonton.pdf.

City of Edmonton. 2015b. *Edmonton's Energy Transition Strategy.* edmonton.ca/city_government /documents/EnergyTransitionStrategy.pdf.

City of Edmonton. 2017. *Climate Change Adaptation: Discussion Papers.* edmonton.ca/city _government/city_vision_and_strategic_plan/climate-change-discussion-papers .aspx.

City of Edmonton. 2018. "Close to 3,400 North American Municipalities Endorse the Edmonton Declaration." *Change for Climate*, June 26. changeforclimate.ca/story/edmonton -declaration.

City of Edmonton. 2019. "Addressing Climate Change: Over Half of Edmontonians Want the City to Do More." Survey results, November 28. changeforclimate.ca/story/survey2019.

The Climate Mobilization. n.d. "Climate Emergency Campaign." Accessed February 27, 2020. theclimatemobilization.org/climate-emergency-campaign.

EPCOR. 2018. "Proposed E. L. Smith Solar Project Received Federal and Provincial Support." News release, November 30. epcor.com/about/news-announcements/Pages/ proposed-el-smith-solar-project-receives-support.aspx.

Explore Edmonton. n.d. "North Saskatchewan River Valley." Accessed April 17, 2020. exploreedmonton.com/attractions-and-experiences/north-saskatchewan-river-valley.

Feroe, Rocky. 2020. "The Deep Solarities." *Just Powers Podcast* (prod. S. Wilson), season 4, episode 4, March 17. justpowers.ca/app/uploads/DS_Feroe_Master_v1.mp3.

Government of Alberta. 2018. Historical Resources Act Approval with Conditions (4941-17- 0008-005). *Culture and Tourism*, September 13. sirepub.edmonton.ca/sirepub/ cache/2/lpxgqwer2atl5mzybhabycdg/837185042020201113537429.PDF.

Government of Alberta. 2020. Bill 001: Critical Infrastructure Defence Act. Second Session, 30th Legislature, 69 Elizabeth II, introduced February 25, 2020. docs.assembly.ab.ca /LADDAR_files/docs/bills/bill/legislature_30/session_2/20200225_bill-001.pdf.

Houle, Rob. 2016. "The Curious Case of the 1908 Enoch Surrender." *Edmonton City as Museum Project*, November 15. citymuseumedmonton.ca/2016/11/15/the-curious-case-of-the-1908-enoch-surrender/.

Issawi, Hamdi. 2019. "Edmonton City Council Declares State of Climate Emergency." *Star*, August 27. thestar.com/edmonton/2019/08/27/edmonton-city-council-declares-state-of-climate-emergency.html.

Jacobs, Beverly, Sylvia McAdam, Alex Neve, and Harsha Walia. 2020. "Settler Governments Are Breaking International Law, Not Wet'suwet'en Hereditary Chiefs, Say 200 Lawyers, Legal Scholars." *Star* (Toronto), February 24. thestar.com/opinion/contributors/2020/02/24/settler-governments-are-breaking-international-law-not-wetsuweten-hereditary-chiefs-and-their-supporters.html.

Johnston, Scott. 2019. "Edmonton to Hit Its Self-imposed Carbon Limit within 8 Years: Report." *Global News*, August 8. globalnews.ca/news/5743781/edmonton-greenhouse-gas-emissions-report-iveson/.

Kelly, Ash, and Hana Mae Nassar. 2020. "Bill 1 Could Spell Jail Time, Huge Fines for Alberta Activists Caught Blocking Critical Infrastructure." *CityNews*, February 26. edmonton.citynews.ca/2020/02/26/bill-1-jail-fines-alberta-blockades/.

Kenney, Jason (@jkenney). 2019. "I agree with these Edmonton residents & the Enoch 1st Nation. The River Valley should be a ribbon of green, free of industrial projects. That's why our govt is ending funding from this solar farm in the Valley & will help create the nearby Big Island Park." Twitter, June 19, 10:30 p.m. twitter.com/jkenney/status/1141533787269918720.

Kenney, Jason. 2020. "Bill 1–The Critical Infrastructure Defence Act." Premiered February 25. YouTube video, 32:38. youtube.com/watch?time_continue=54&v=wC5ykzYg8Qw&feature=emb_logo.

Lindsay, Don. 2020. "Teck Withdraws Regulatory Application for Frontier Project." Teck news release, February 23. teck.com/news/news-releases/2020/teck-withdraws-regulatory-application-for-frontier-project.

Metz, Emily. 2019. "City of Edmonton Declares Climate Emergency." *Global News*, August 27. globalnews.ca/news/5821850/edmonton-climate-emergency-greenhouse-gas-emissions/.

Mirrorland Collective. 2020. "A Big Pile of Glitch: A Manifesto for Feminist Solarity." *Just Powers*. justpowers.ca/projects/feminist-solarities/.

Paris Agreement, United Nations Framework Convention on Climate Change, December 5, 2015. unfccc.int/process-and-meetings/the-paris-agreement/the-paris-agreement.

Richmond, Charles. 2020. "The Deep Solarities." *Just Powers Podcast* (prod. S. Wilson), season 4, episode 5, March 17. justpowers.ca/app/uploads/DS_Richmond_Master_v1.mp3.

Rieger, Sarah. 2020. "Teck Withdraws Application for $20B Frontier Oilsands Mine." *CBC News*, February 23. cbc.ca/news/canada/calgary/teck-frontier-1.5473370.

Sharphead, Cody. 2020. "The Deep Solarities." *Just Powers Podcast* (prod. S. Wilson), season 4, episode 3, March 17. justpowers.ca/app/uploads/DS_Sharphead_Master_v1.mp3.

Simpson, Audra. 2014. "The Chiefs Two Bodies: Theresa Spence and the Gender of Settler Sovereignty." Lecture presented at Unsettling Conversations, Unmaking Racisms and Colonialisms, R.A.C.E. Network's 14th Annual Critical Race and Anticolonial Studies Conference, University of Alberta, Edmonton, AB. October. vimeo.com/110948627.

Teck. 2020. "Teck Announces Purchase of SunMine Solar Energy Facility." News release, January 15. teck.com/news/news-releases/2020/teck-announces-purchase-of-sunmine -solar-energy-facility.

Theodorakidis, Alexandra. 2015. "Bill C-51, Freedom of Assembly and Canadians' Ability to Protest." *CJFE: Canadian Journalists for Free Expression*, June 27. cjfe.org/bill_c_51 _freedom_of_assembly_and_canadians_ability_to_protest.

Wilson, Sheena. 2017. "Gender." In *Fueling Culture: 101 Words for Energy and Environment*, edited by Imre Szeman, Jennifer Wenzel, and Patricia Yaeger, 174–75. New York: Fordham University Press.

Wilson, Sheena. 2020. "Petro-Misogyny: Don't be Duped (Again)." *Deep Energy Literacy*, April 21. deepenergyliteracy.csj.ualberta.ca/2020/04/21/misogyny/.

Rhys Williams

Turning toward the Sun:
The Solarity and Singularity of New Food

As Nicole Starosielski (2019) claims, "agriculture is a culture of solarity:" a set of embodied practices, mediations, and infrastructures, entangled in labor, place, climate, natural rhythms, and community. The dominant mode of global industrial agriculture is beset with crises of declining returns, increasing use of land and water and energy from fossil-fueled inputs, and is one of the main systems driving climate change. In this hour of need, a cluster of New Food companies are emerging that promise to revolutionize food production. Broadly speaking, their solution involves decoupling from farming animals and, for the more radical examples, from agriculture itself, recasting food in terms of its composition rather than its origin, and producing it using renewable energy. All of which relies on intensive investment in technology and infrastructure, and a complete transformation of the global food regime as we know it.

The key innovation that underpins the promised transformation is precision fermentation (PF). PF is the name given to the combination of fermentation and precision biology (PB). In PB, "the coming together of modern information technologies . . . with modern biotechnologies" (Hinds 2020) means that, in the words of booster think

The South Atlantic Quarterly 120:1, January 2021
DOI 10.1215/00382876-8795791   © 2021 Duke University Press

tank RethinkX's "Sector Disruption Report" (written by technology and finance gurus Tony Seba and Catherine Tubb), "biology has undergone a conceptual shift by becoming an engineering discipline. . . . Just like software developers, synthetic biologists can engineer biology" (Seba and Tubb 2019: 17). In this particular case, we're talking about the capacity to rewrite or imprint new genetic code into simple microorganisms, so that during the PF process "the microbes then act as highly efficient factories that consume specific inputs and spit out desired outputs" (Hinds 2020). Although more established New Food companies like Impossible Foods or Beyond Meat produce plant-based meats and remain tied to crop farming, more radical companies like Air Protein and Solar Foods use PF to produce proteins from "thin air" and claim to be completely decoupled from agriculture. The success of the latter will in turn make emerging cultured meat companies significantly more viable (for example: Memphis Meats; Shiok Meats; Finless Foods). Currently, the costly feedstock required by cultured cells prices them out of the market; this feedstock could be produced far more cheaply with scaled PF technology.

PF food production essentially takes place in a factory that requires water, electricity, microbes, and CO_2. The water and electricity are used to produce hydrogen, which is one of the costliest and most energy-intensive parts of the process. Couple it with renewable energy, however, particularly solar, and the entire process appears almost demiurgic in its capacity to create food from the elements. Just as industrial agriculture and fossil fuels go hand in hand, so do New Food and solar: both narratively and infrastructurally, they form a new and shiny capitalist food production assemblage—a new solarity—for the twenty-first century. Their emerging synergy is no accident: both technologies were developed together as infrastructural solutions to inhospitable environments, to allow humans to survive where there is *only infrastructure* between life and death. NASA was initially responsible for developing into practical use both solar panels and the single-celled organisms (hydrogenotrophs) that are used to convert CO_2 into food without the need for photosynthesis. Solar Foods is currently in partnership with the European Space Agency to develop food production for a Mars mission (ESA 2019). These two technologies—solar and PF—combine to create a closed carbon-cycle, a total life-support system, a spaceship. These are the ecological solutions now being offered to ameliorate our rendering inhospitable of the planet. The basic logic of the solution is that of apparently uncoupling from the Earth and its ecosystems in order to save both it and ourselves.

The technological developments in PF that underpin New Food are accompanied by a conceptual shift, necessary to make narrative sense of the new relationship to the world that the technologies embody. This shift lies in understanding food as a composition of elements as opposed to something that takes its identity from its origin, which is to say, to thinking of food as information that can be built up from its component parts—"food as software" as Seba and Tubb (2019) would have it—rather than broken down from its traditional sources (animals and plants). This move reframes a cow, for example, as a poorly designed machine or "chassis" for the production of particular commodities, one that can and should be "unbundled" (14), and the new technology of PF can be utilized to produce the specific proteins desired at a far higher efficiency and at less cost. This same basic point applies to all industrial agriculture, crops as well as animals.

To move the basic unit of production from macroorganisms to microorganisms like this is to uncouple food production further from the ecosystemic modes and relations that we call nature, and to reground it in those infrastructural modes and relations of production we call capital. The industrial agricultural system strove to render animals and plants as capital in so far as it was able, but the holistic system that is the animal or plant itself, and its relationality to the ecosystems around it—however dominated and shaped by capital—remains partially beyond the reach of capital, a source of structural disagreement that resists legibility and control. If the new food system does away with plant and animal, it also does away with that specific form of resistance, rendering the unbundled macroorganisminto categories of producible commodity, and abjecting the rest.

In the heralded "domestication of micro-organisms" (14), we can hear a secondary promise that these microorganisms will be finally removed from the realm of the political. The shift to domestication of microorganisms is driven by the desire to reduce mediation and render smooth the structural disagreement that was manifest between the nonhuman world and the infrastructure of industrial agriculture. Seba and Tubb (14; emphasis mine) are clear on the benefits here: "We can replace an extravagantly inefficient system that requires enormous quantities of inputs and produces huge amounts of waste with one that is precise, targeted, and *tractable*." This is not to say that disagreement will not emerge in this new distribution of things, possibly with extreme and unforeseen consequences—if food is to be software, then it will doubtless have a few bugs after all—but it is difficult to shake the sense of the ground of politics receding here.

The plausibility of New Food's rise depends upon the way it intersects with and mobilizes dominant discourses of power. One of the beliefs that provides the basic structure for such contemporary dominant discourses is that of continuing technological progress. Seba and Tubb premise their narrative of the inevitable rise of PF companies and the collapse of industrial agriculture on "the reality of fast-paced, technology-adoption S-curves" (3). S-curves are, however, a historically contingent narrative structure, rather than a reality. Their roots—the "general idea of exponential accelerating change" (Broderick 2012: 21)—go back to a science fiction author, Robert A. Heinlein, who in 1952 projected what he called the "curve of human achievement" to "go on indefinitely with *increasing* steepness" (21). That the statement of this belief would emerge simultaneously with the Great Acceleration is no accident. From here the idea develops until popularized most famously by Vernor Vinge, who proposed that "exponentially accelerating science and technology are rushing us into a Singularity" (20). This proposal is premised on the idea that "technological time" is a "series of upwardly accelerating logistical S-curves, each supplanting the one before as it flattens out" (20). This is precisely the shape of the narrative of New Food as told by its companies and its proselytizers: a bottleneck into which emerges the slow tail of a new technological S-curve, bound to overtake and disrupt what came before, as natural and inevitable as evolution itself.

The narrative of the Singularity and the desires and values it harnesses are far older than its current digital flavor. It is at bottom a narrative about the desire to be freed from the burdens of history and necessity, from responsibility and care for others, and from frustrations and restrictions on agency: a narrative of liberation from mediation. At moments of exuberance and faith in the power of technology and human ingenuity, especially at times fueled by a dramatic new energy input, the imaginary of the Singularity achieves its escape velocity. We see this in the nineteenth-century odes to steam (Malm 2015); in the dazzling new horizons of Futurama in the 1930s and 1940s (General Motors 1940); during the Great Acceleration and the science fiction of the 1950s and 1960s (Canavan 2014); and now with solar power and New Food (Williams 2019). The Singularity's desire for freedom from place, climate, natural rhythms, and community is the desire to reject precisely that which allows our existence as precarious and relational beings. The mechanism for this desire is to create a temporary mirage of liberation through containment, through uncoupling from the world.

Containment in sterile fermentation tanks is the condition of possibility of increased control over protein production. In this space, protein production is allegedly rendered completely malleable, transformable, a matter of matter, stripped of history or situatedness. Via containment comes the promise of infinity:

Free to design molecules to any specification we desire, the only constraint will be the confines of the human imagination. Each ingredient will serve a specific purpose, allowing us to create foods with the exact attributes we desire in terms of nutritional profile, structure, taste, texture, and functional qualities. Virtually limitless inputs will, therefore, spawn virtually limitless outputs. (Seba and Tubb 2019: 14)

Further, PF food production is uncoupled from place and climate. Seba and Tubb (21) proclaim: "Food production will no longer be at the mercy of geography, or of extreme price, quality, and volume fluctuations due to climate, seasons, disease, epidemics, geopolitical restrictions, or exchange-rate volatility," and each company stresses the same benefits. Air Protein (2019) talks about "allowing farms to expand vertically with geographic flexibility" and production happening "independently of weather conditions or seasons." The CGI video showing a Solar Foods production facility is appropriately abstract, nothing more than a square of solar panels in a blank desert-like space, with a fermentation facility in the middle: a no-place, sprung fully formed from the brain of capital and locatable anywhere. Seba and Tubb (2019: 20) envisage food production "shifting from large, remote, agricultural areas to smaller, easily accessible, urban areas" with a new decentralized network of local production and consumption that also demonstrates its sympathy with narratives of solar futures such as Hermann Scheer's (2004). It also uncouples from the dictates of seasonal productive rhythms: New Food's "probiotic production process is ready for harvest in hours" rather than "months to go from seed to harvest," and if you want to make more, you just build another fermentation tank (Parletta 2019).

The driving force and persuasive power behind the New Food narrative and its Ur-narrative, the Singularity, lies in yoking together the concepts of total containment on the one hand, and pure, impossible liberation on the other. The more complete the liberation, the more total the act of technological containment, or uncoupling from the world, required. The more total the uncoupling, the more sublime the liberation promised: not only from livestock, and not only from crop farming, but ultimately from the restrictions of the natural world itself. This is why solar and New Food make such a potent combination in this time of climate collapse. They are able to present themselves as ecologically friendly solutions precisely because they are cut off from ecosystems, forming a closed loop of their own.

Crucially for its narrative success, however, New Food doesn't present itself as simply a radical rupture with the familiar. It claims to be a heroic solution to the contemporary ecological crisis, with particular emphasis on a huge decrease in land, water, and fossil fuel use. Its key move is to reframe what was previously an image of human domination and power—large agricultural machinery like the combine harvester, rows upon rows of harvested crops stretching into the distance, the dust and exhaust kicked up by machines—as an image of sprawl and inefficiency. However, when presenting their radical production techniques, New Food companies make consistent rhetorical gestures towards tradition, purity, and nature, while the clean aesthetics and images of sanitized nature that fill the websites corroborate the story. Solar Foods's (2019) "bioprocess . . . is natural, with a fully natural fermentation process" utilizing microbes collected from "pure Finnish nature" to produce the "purest . . . protein in the world"; Air Protein's (2019) process is "similar to making yogurt or beer"; Impossible Foods (2020) make them "the way Belgian beer is made." It's new but reassuringly linked with the past, a natural next step. It also offers a means of continuing to enjoy food as we currently do (at least, as those in affluent countries do) rather than demanding we change our consumer habits. The narrative allows for the porting of present desires, values, and their associated practices from the present into the future, with minimal fuss. As Impossible Foods (2020) tells it:

> We've been eating meat since we lived in caves. And today, some of our most magical moments together happen around meat: Weekend barbecues. Midnight fast-food runs. Hot dogs at the ballpark. Those moments are special, and we never want them to end. But using animals to make meat is a prehistoric and destructive technology. We're making meat from plants so that we never have to use animals again. That way, we can eat all the meat we want, for as long as we want. And save the best planet in the known universe.

The consumer doesn't have to do a thing except choose a better version of what they already consume, and this is only the beginning. Once PF kicks in, and cultured meats make their way to the supermarket, even the conflicting tug of real meat will be dispensed with. Writing this in April 2020, in the midst of the COVID-19 lockdown in the UK, it is hard not to think how receptive the post-COVID world will be to the narrative of New Food, and even more unreceptive to ecological narratives that frame contamination as a positive value (Alaimo 2016). The inevitable demand for the greater securitization of food, the questioning of supply chains, the calls to shut down wet markets, the links made between industrial agriculture and the disease, and the barely submersed undercurrent of racist purity discourses, are all answered perfectly by New Food's narrative. In fact, the Good Food Institute (a pro–New Food think tank) swiftly published an article in *Wired* making the link (Specht 2020).

New Food is also pitched as the hero of a second narrative, that of profit-making potential. Here, again, a situation is identified—declining rates of profit, more input required for less output—in response to which New Food promises enormous rewards for the bold investor. In terms of translating New Food from a narrative promise into a dominant infrastructural force, both this narrative and that for a general audience are important. But the presence of the economic narrative is noteworthy. It takes the present state of things and translates it through comprehensible and achievable actions for existing powerful actors—profit-driven investment in and construction of new infrastructures. These actions lead to a future underpinned by those infrastructures, in which they will be immense sources of value and wealth creation. Here, again, the narrative acts as a vehicle for porting value and desire from the present into the future. In a virtuous circle of narrative influence, a third narrative then becomes possible, where wealthy investors are cast as heroes just for investing in New Food, while politicians are heroes if they get out of the way or simply legislate favorably. This sits atop a fourth, more fundamental, set of neoliberal narrative coordinates about entrepreneurial drive and ambition and a lumbering government apparatus (Szeman 2015). Narrative begets narrative, enrolling actors into its realization by casting them as heroes, making itself material by providing the value system by which to translate actions and money into infrastructure, and so into concrete politics—to world-build. As the Financial Times puts it: "The desire to save the planet is not the only force helping to advance entrepreneurship. A growing number of investors, from venture capital firms to food and agriculture corporations, are eyeing potential returns from funding innovations" (Murray 2019). Or, in the words of Impossible Foods (2020), winners of the UN's 2018 Champion of the Earth Award: "Our mission demands relentless growth every year."

Consider this as a figure for New Foods: An astronaut in a bulky white spacesuit stands in a parking lot, a golden visor concealing his face. He kneels to study a plant growing in a crack in the tarmac. We hear him speak through his radio: "There's life." He moves through different everyday scenes—sitting at a bus stop, passing a fruit stall, lying in a park—always bathed in sunshine, marveling at the world he sees: "The colors, the beauty, the movement, it looks like a living, breathing organism. It's so beautiful here. I've dreamt of this planet for so long. This is paradise." Finally, in the woods, with the sun setting behind him, he releases his helmet visor to uncover his face and takes his first deep breath of native air. The video closes with text: "We're on a Mission (and it's not to Mars)."

This is "The Return," a short promotional video for Impossible Foods. In it we can read a figure of New Food's self-image, and perhaps some unintended truths. The narrative deploys the typical science fiction affordance of estrangement, in which we are placed at a distance from our world and asked to see it anew. The premise is a simple one, based on lampooning competing techno-utopian plans to save humanity by going to Mars when, in the words of the astronaut, "everything is here." The narrative is a blow struck in a narrow battle over the right direction for techno-utopianism in contemporary culture. The astronaut is the pioneering spirit, representative of the heroism of technoscience and the reconnection to, and restoration of, nature that the technologies of New Food promise.

Yet the spacesuit—the vehicle for the narrative and the mechanism of estrangement—is a condensed image of a highly infrastructurally mediated relation to the world, one that casts the environment as hostile, in which the world is filtered, transformed, and blocked for survival's sake. In this narrative, we have to have our sensibilities rerouted through the estrangement of technoscience, and our necessities (food, air, heat, waste recycling) ameliorated by it, before we can appreciate the beauty of the Earth. This is not an image of the future, but of the present for those privileged enough to be removed from the necessity of an intimate laboring relation to nature, who have their needs met by infrastructures. Once released from necessity, we are free to reestablish our relations to nature on new, distant, and purely aesthetic grounds (a relation that obviously has a long and deeply racialized history, particularly in the US). The real fantasy here is that one day, with enough science and technology and human ingenuity, the visor will open: mediation will be rendered unnecessary, and we will no longer need the suit. Ironically, it is a narrative that valorizes technoscience through a fantasy of escape from its entanglements: through greater investment in the narrative thrust of science fiction, it promises us Eden.

What might be the response be to such a promise? While many argue now, rightly, for the necessity of working to make better mediations, more livable relations, of paying more attention to the world, of hearing and understanding it, and bending our practices and infrastructures towards it, this is not the thrust of dominant cultural narratives, nor does this sort of work resonate well with many dominant, underlying structures of belief, required to render narratives plausible. In addition, such work is so painstaking and complex it doesn't lend itself well to narrative, or at least not to easily digestible and gripping tales of heroism that turn on dramatic conflict and sweeping solution. The fundamental narrative issue is that such ecological work is not a solution at all, but rather an ongoing and necessary response to a permanent and changing condition of mediation. Techno-utopian imaginaries like that of New Food, on the other hand, tend to deploy symbolic resolutions to real contradictions (*pace* Fredric Jameson) via what science fiction critics would call a "novum"—an imaginary transformational infrastructure that does not exist. Or, at least, not outside the lab and not at the scale required. These promised infrastructures are typically present in narrative far more than in the world, present more as form than as material, as aesthetic more than technic, suturing past and present to a future with which they are otherwise irreconcilable. This is a science fictional mode of narrative-making, where narrative progression (from here to there) is conveyed through technology's capacity to transform. On the one hand, these future imaginaries provide a justification for continued commitment to business as usual, via a renewal of the old Promethean tale of human ingenuity and overcoming. On the other hand, they shift the present towards that future. They render that sequence of events between here and there plausible by tying it to existing mechanisms and narratives, and consequently reroute those mechanisms— entrepreneurial activity, funding streams, scientific and engineering research and development, media attention, political and public support— via the transformational technology and towards the realization of the initial narrative promise. New Food, and thus the consistency of human/nonhuman relations, is now at a critical juncture. The conditions are ripe for its narrative promise to be heard, and its more mundane vanguard is already an infrastructural and economic reality, preparing the ground for the more radical of its kind that follow.

To push these ideas to their limit: New Food's ultimate promise of infinite liberation (of proteins, in this case) should be taken with a pinch of salt. As is usual under capitalism, while infinite variety is promised, the result is often endless and superficial false novelty. New Food is persuasive because it meshes so seamlessly with the present conditions of production and consumption, and it is this same quality that ensures a lack of genuine difference in a future where it is successful. What New Food produces is not difference but undifferentiated substrate. The form of its emergence into the world is dictated by consumer categories and circuits of capital (the Burger King Impossible™ Whopper; White Castle Impossible™ Sliders; Little Caesars Impossible™ Supreme Pizza; and so on). Whereas it took thousands of years of agriculture and animal husbandry and the evolution of food culture to break down the given of animals and plants and creatively work it up through entire historical cultures of food gathering and preparation into sustenance entwined with ritual and meaning, New Food provides no resistance with which culture must work, no mediation that demands invention. Instead, New Food provides a perfect pap, pure content that takes on any form of capital's invention, a perfect product, pure commodity from the moment of its inception.

At this point, a certain purity of logic makes it tempting to argue that New Food is not only fundamentally continuous with industrial agribusiness but represents a triumph of its basic logic and ontology that was hitherto unachievable: a stripping away of complicating relations, a purer emergence or unfolding of a solipsistic mode of relating to the world. Rather than constituting a difference, it promises instead to abject all difference. The strong version of this claim—there is no essential difference between New Food and industrial agriculture—has the benefit of making clear the continuities that might otherwise be obscured by the transformative discourses associated with PF technology. However, it is also absurd to disregard the meaningful differences, in the proposed elimination of animal cruelty, soil depletion, and agricultural CO_2 emissions.

Like all good science fictional novums, New Food balances on a knife-edge between threat and promise. This article seeks to clarify the threat, in the midst of the chorus of promise. The long-term ecological orientation embodied by New Food, for all the initial good it promises and may well deliver—is not that of developing a better relationship with nonhuman nature but of separating, for the alleged good of both parties. An amicable divorce, but one in which the terms of the settlement—you have that, we'll have this, and we'll share the earth—will not withstand the basic drive of growth, progress, assimilation, and control that characterizes the Singularity logic of New Food. The contemporary foil for solar and New Food is the grim opening of *Blade Runner 2049* (Villeneuve 2017), where the ground is entirely covered by solar panels straining to generate power in the gloom, and the food is produced by "synthetic farming" in vile hinterland laboratories. The uncoupling from the ecosystem is the same, only here we find the abjection that is the dark twin of estrangement. Ideas of rewilding freed-up land might be naïve: this narrative and infrastructural thrust points instead to Spaceship Earth, powered by a Dyson sphere, fully securitized, controlled, and geo-engineered.

To be clear: the natural world presents as an enormous variety, but within that variety there are linkages that both limit and generate—interdependencies between flora and fauna, insect, plant, mineral, and animal.

These are contingencies in that they depend upon each other, and their existence is uncertain and changeable to the extent of that dependence. This contingency is what makes the future a place of genuine novelty. New Food presents us with the mirage of freedom from that contingency—the world can be made anything, all the time, limitless and free. But it will exchange one complex contingency for another simpler one—a total reliance on capital and its infrastructures, powered directly by the sun, fed directly by the sun—and genuine freedom for virtual freedom in containment. Eventually, the natural world in its entirety may come to be seen as simply an inefficient and outmoded system of production, or worse, an indefensible source of contamination. Consider this as we gradually pull on our spacesuit; be sure to savor the air before the visor comes down.

Note

The author would like to thank Darin Barney, Alex Campbell, Chris Maughan, Imre Szeman, and Briony Wickes for their generosity, their discussions, and their comments; Rebecca Goldie for her support; and the University of Glasgow for providing the time and means to work on this article.

References

Air Protein. 2019. www.airprotein.com.

Alaimo, Stacy. 2016. *Exposed: Environmental Politics and Pleasures in Posthuman Times*. Minneapolis: University of Minnesota Press.

Broderick, Damien. 2012. "Terrible Angels: The Singularity and Science Fiction." *Journal of Consciousness Studies* 19, no. 1–2: 20–41.

Canavan, Gerry. 2014. "Retrofutures and Petrofutures: Oil, Scarcity, Limit." In *Oil Culture*, edited by Ross Barrett and Daniel Worden, 331–49. Minneapolis: University of Minnesota Press.

ESA. n.d. "Food out of the Thin Air." www.esa.int/Applications/Telecommunications_Integrated_Applications/Technology_Transfer/Food_out_of_the_thin_air (Accessed 23 April 23, 2020).

General Motors. 1940. *To New Horizons*. archive.org/details/ToNewHor1940.

Hinds, Taylor. 2020. "Precision Fermentation: What Exactly Is It?" *RethinkX Blog* (blog), January 8. blog.rethinkx.com/precision-fermentation-what-exactly-is-it/.

Impossible Foods. 2020. impossiblefoods.com.

Malm, Andreas. 2015. *Fossil Capital: The Rise of Steam-Power and the Roots of Global Warming*. London: Verso Books.

Murray, Sarah. 2019. "Investors Develop a Taste for Radical Thinking on Food." *Financial Times*, March 11. www.ft.com/content/6674f2b6-d165-11e8-9a3c-5d5eac8f1ab4.

Parletta, Natalie. 2019. "Meatless Meat Is on the Horizon—You'll Be Surprised Where It Comes From." *Forbes*, August 4. www.forbes.com/sites/natalieparletta/2019/08/04/meatless-meat-is-on-the-horizon--youll-be-surprised-where-it-comes-from/#3133c6a87ed2.

Scheer, Hermann. 2004. *The Solar Economy: Renewable Energy for a Sustainable Global Future.* London: Earthscan.

Seba, Tony, and Catherine Tubb. 2019. *Rethinking Food and Agriculture 2020–2030: The Second Domestication of Plants and Animals, the Disruption of the Cow, and the Collapse of Industrial Livestock Farming.* RethinkX. www.rethinkx.com/food-and-agriculture.

Specht, Liz. 2020. "Modernizing Meat Production Will Help Us Avoid Pandemics." *Wired,* March 13. www.wired.com/story/opinion-modernizing-meat-production-will-help-us -avoid-pandemics/.

Starosielski, Nicole. 2019. "Harvesting Sunlight" Keynote presented at the conference "After Oil 2: Solarities," Montreal, QC, May 23.

Szeman, Imre. 2015. "Entrepreneurship as the New Common Sense." *South Atlantic Quarterly* 114, no. 3: 471–90. doi.org/10.1215/00382876-3130701.

Villeneuve, Denis, dir. 2017. *Blade Runner 2049.* Burbank, CA: Warner Bros. Pictures.

Williams, Rhys. 2019. "'This Shining Confluence of Magic and Technology': Solarpunk, Energy Imaginaries, and the Infrastructures of Solarity." *Open Library of the Humanities* 5, no. 1: 60. olh.openlibhums.org/articles/10.16995/olh.329/.

Gökçe Günel

Leapfrogging to Solar

In July 2018, the renowned American business magazine *Forbes* ran an article on solar power in Ghana titled, "Why This 48-Year-Old Woman is Building Ghana's Biggest Solar Farm." Salma Okonkwo, the chief executive officer of UBI Group, a company that specializes in downstream petroleum operations, was the person behind the project (Sorvino 2018). Blue Power Energy was planned to open in March 2019, with a capacity of one hundred megawatts. In a country where the electricity grid's total volume was about five thousand megawatts, a one hundred–megawatt addition was noteworthy. In November 2019, the International Energy Agency featured Okonkwo in an article titled "Seven Women Entrepreneurs of Solar Energy" (IEA 2019). Other news sites, such as *Ventures Africa* and *AllAfrica*, echoed the news, and celebrated Salma Okonkwo's success (Durodola 2018; Kiunguyu 2018). Along with another much-publicized plant called the Nzema Solar Power Station, located in the western part of Ghana and built by a British company called Blue Energy, Okonkwo's solar power station promised to revolutionize Ghana's energy future.

During a visit to Ghana's Energy Commission in December 2019, however, Kofi, an electrical engineer who had been working with the

The South Atlantic Quarterly 120:1, January 2021
DOI 10.1215/00382876-8795803 © 2021 Duke University Press

regulator body for more than a decade, was quick to tell me that neither of these two plants existed. Kofi was in charge of the country's renewable energy vision, aiming at 10 percent renewable energy on the grid by 2030, as espoused by the Paris Agreement, and he administered the bureaucratic processes for registering renewables.[1] He told me how the Nzema Solar Power Station was the first solar power station to receive permission to start construction from the Energy Commission in the period 2012–16. Due to its location on the western coast of Ghana, the plant took up too much farmland, and the company eventually gave up negotiating with the chiefs that governed the region. It was never built. Yet the plant advertised by *Forbes* had a more perplexing story, according to Kofi. It had never even been registered. The commission's website featured all of the documents related to renewable energy projects, and there was no trace of Salma Okonkwo in these files. Kofi could not understand how a respectable publication like *Forbes* would publish an article on a one hundred–megawatt plant without fact-checking. "If you end up finding the plant, I'll pay for your taxi ride there," he joked. A one hundred–megawatt photovoltaic power plant would require a large site; therefore, the station would be difficult if not impossible to conceal. Some days after our meeting in his office, I texted him on WhatsApp to ask if we should really try to find the plant in northern Ghana, "I don't know where we will be going to, NoWhere [*sic*]," he wrote back. His colleagues at other institutions in Ghana, such as Electricity Company of Ghana (ECG) and GridCo, confirmed that the plant was practically fictional and summed it up as "fake news." Despite my attempts, I could not secure a meeting with Salma Okonkwo herself.

Why exactly has the Okonkwo story attracted attention in the international media? Discussions of energy in sub-Saharan Africa tend to focus on leapfrogging, theorizing how the region might be able to avoid carbon-intensive fuels, such as oil, coal, and natural gas, and directly start using renewable energy infrastructures (see, for instance, Blimpo et al. 2017). In an article on electricity in the global South, anthropologist Akhil Gupta (2015: 564) suggests, for instance, that "the real question is how the quality of people's lives in the global South can be improved, not by emulating the destructive patterns of the global North, but by pursuing new, sustainable developmental paths that have not yet been taken. There is a real opportunity to leapfrog the carbon economy and life on the grid to establish future-oriented, sustainable ways of living." Yet such attractive, future-oriented, and global visions do not always take into account the contexts in which these renewable energy power stations will be built and who exactly they will benefit.[2] This is perhaps one of the reasons why the story of the solar power plant was so attractive to *Forbes*: it indicated a green energy future, led by an African woman, whose work evidenced that leapfrogging might be possible, even though this belief was not necessarily matched by transformations happening on the ground.

In the year 1997, hydroelectric power stations produced all of Ghana's electricity (Awopone et al. 2017). Yet increasing power demand and lower water levels in hydropower stations due to climate change have resulted in an energy grid that now relies on thermal power stations for two-thirds of its electricity production (see also Silver 2015). Energy professionals in Ghana told me that institutions like the World Bank merely paid lip service to the imagination of a renewable energy future, imposing global topics of significance onto the global South, even though these topics did not have much domestic weight. In the meantime, the World Bank invested in fossil fuel infrastructures, such as the Sankofa Gas Reservoir, an offshore gas exploration project in western Ghana (World Bank 2015). Energy infrastructure in Ghana did not necessarily change along a singular, linear route of transition to renewable energy, or even along a singular pathway. Rather it proceeded based on binding bilateral contracts with power producers. In a context where electricity was scarce and difficult to access, transformations did not happen along a coherent premeditated strategic path but responded to seemingly erratic, ad hoc, and black-boxed market logics, resulting in higher and higher rates for industrial, institutional, and residential consumers. Given that energy demand is expected to increase in other parts of sub-Saharan Africa, this type of transformation is likely to occur in other parts of the continent as well.

While the energy portfolio of Ghana indicated a shift from hydropower to thermal stations, some individuals and institutions in the country with access to upfront capital have invested in rooftop solar panels. Affluent individuals, such as Fadi and Solomon, whose experiences I will share below, began to see rooftop solar panels as a way to decrease their energy costs and their reliance on an unsteady grid, causing this new technology to emerge as a status symbol in Ghana. Many institutions, such as the Australian High Commission, CAL Bank, or the Kasapreko Bottling Plant in Accra, also followed this trend.[3] As these consumers minimized their debt to the Ghanaian national grid, electricity distributors in the country had trouble paying independent power producers and balancing their books. Eventually, they responded by increasing rates for consumers who most likely could not afford the upfront costs of rooftop solar panels, and who had no choice but to remain on the grid. The construction of rooftop solar panels by the high-paying facilities had the unexpected and undesirable effect of making electricity consumption a heavier financial burden on consumers across the country, increasing inequality.[4]

Finance professionals both inside and outside Ghana understood such inequality as a business problem, which could be resolved through the development of innovative green financing tools, most importantly green loans. Their approach followed a larger trend towards "making finance flows consistent with a pathway towards low greenhouse gas emissions and climate-resilient development," as predicated by article 2 of the Paris Agreement. By allowing bankable middle-class Ghanaians to apply for low-interest, three-to-five-year loans to build rooftop solar panels, financial institutions such as CAL Bank and Stanbic Bank aspired to expand the populations who benefit from already available rooftop solar technologies. One aid worker told me how climate change and energy issues were no longer related to having access to the right technological solutions, but rather necessitated being able to rely on and reproduce the right business models. Anthropologist Jamie Cross (2019: 48) points to how "selling solar power to people living in chronic energy poverty presents itself as an ethical economic utopia: the opportunity to express care for others and the environment at the same time as fulfilling a fiduciary duty of care to investors and shareholders." A wider look at the solar energy landscape in Ghana shows that it is not necessarily the solar panels themselves that fulfill this "fiduciary duty of care," but rather green loans, which might facilitate the adoption of solar power and other energy efficiency technologies among bankable consumers. This is one of the reasons why the finance professionals I met in Ghana perceived themselves to be fundamental to facilitating a shift to a renewable energy future in the global South and hoped for educational programs that would teach Ghanaians the value of efficient energy use. Confronting the tangible effects of climate change, those educated Ghanaians would then find more value in the green financing programs their banks offered.

It is important to track and document the transformations in electricity generation at the national and global scale, but an exclusive focus on indicators at these scales runs the risk of rendering the intricacies of new energy infrastructure invisible. In a context where larger trends favor thermal infrastructure, what kinds of social and financial relations do small-scale renewable energy projects, such as rooftop solar panels, generate? Thinking in particular about solar power, the introduction to this volume suggests: "The belief that solar can meet our energy needs *and* take care of humanity, and can accomplish both via technology alone, is at the heart of the promise of solar. It is a belief that foregrounds the sun's abundance and the human ingenuity required to harness it, while relegating to the shadows the very real challenges of producing energy transition on a global scale" (Szeman and Barney 2021). Beyond celebrating the arrival of clean technologies as agents of climate change mitigation, the literature on renewable energy pauses to probe how and if these new technologies might contribute to the production of just societies and asks if renewable energy power stations might give rise to opportunities for social change (Boyer 2019; Günel 2019; Howe 2019). An analysis of rooftop solar panels in Ghana demonstrates how the adoption of these new technologies facilitates a kind of leapfrogging to renewable energy for a select few. Surprisingly, such pragmatic, seemingly ethical, and environmentally conscious choices render electricity consumption even more out of reach for those who rely on the grid for their electricity needs.[5]

Business Class Passengers

In the 1980s, Fadi's father purchased an island that faced on one side the Atlantic Ocean and on the other the Volta River, a unique spot among an archipelago of small islands scattered at the river's mouth. During the weekends, his family would entertain their many guests atop this whimsical patch of sand, carrying essentials on a high-speed boat from the nearest fishing village, Ada. For decades, they transported diesel for their generators too, but this arrangement changed when Fadi's savvy cousin took on the responsibility of cladding the house in rooftop solar panels. Their initial investment of 187,000 Ghanaian cedi paid off, in only a few years.[6] Fadi's family had been importing car parts and luxury vehicles to Ghana since the mid-twentieth century and did not seem to have any trouble financing the project. The Porsche SUV that Fadi drove was one of the models that they had recently introduced to the Ghanaian market and had so far been doing well. As we discussed solar power production, Fadi's cousin singled out their batteries as the technology most in need of advancement, an argument put forward by many in the renewable energy sector. They kindly offered me a beer that had been cooled in a solar-powered fridge and told me to enjoy myself alongside their other visitors on this beautiful Sunday afternoon.

"Rooftop solar is a status symbol, like owning a luxurious car," Belinda, a young energy professional with the Energy Commission, argued when I told her about my experience visiting Fadi's island near Ada, and confirmed that most rooftop solar power stations were self-financed. Moreover, this was not only the case for those with homes off the grid, such as Fadi's. Even when they did have easy access to the grid, people with resources opted for rooftop solar panels. I mentioned to Belinda that I had also visited Solomon, a well-renowned man about town Belinda had heard of but never met, who had outfitted his home and office with solar panels, even though he lived in central Accra near the American Embassy with full access to ECG's electricity supply. Solomon had spent most of his life in Washington, DC, working with various international organizations, such as the World Bank. Later in his life, he moved back home to Accra and took on responsibilities as a policy advisor while writing informative but hard-to-come-by coffee-table books about Ghana. A well-traveled and well-connected man, Solomon was proud of his rooftop solar project, and used a small monitor to check on his production levels throughout the day. Nearly half of his electricity was produced by the seven-kilowatt panels he placed on his roof. When I asked what prompted him to install the panels, he referred to the persistent electric power outages that Ghanaians had been experiencing, christened *dumsor,* meaning off and on.

During the final *dumsor* episode from 2012 to 2015 in Ghana, power for industries and homes was out for twenty-four hours at a time and turned back on for only twelve-hour periods (Ahlijah 2017). Forced to diversify the country's energy portfolio, Electricity Company of Ghana signed forty-three power purchase agreements with different vendors, altering the country's energy production portfolio significantly. Independent power producers from countries such as Turkey, China, and the United Arab Emirates began generating electricity, mainly in power stations in the cities of Tema and Takoradi. Many of these independent power producers entered the Ghanaian market on what the industry defined as "take-or-pay contracts," where the Electricity Company of Ghana either bought electricity from a producer or paid the producer a penalty for electricity it did not use. "We over-solved the problem," one energy professional who had been a junior member of a team at the Energy Commission, a regulator body responsible for electricity licensing, reflected back on the crisis. In 2020, five years after the end of the electricity crisis, the country had an installed power generation capacity of about five thousand megawatts, although peak demand rarely exceeded twenty-eight hundred megawatts. Faced with energy excess, executives at the Electricity Company of Ghana felt that they had no choice but to increase rates for consumers.

The rate increases led many individuals, institutions, and industrial facilities to rethink their electricity consumption patterns. As a response, many built rooftop solar panels, cutting their reliance on the Electricity Company of Ghana by half. A green finance specialist I spoke with used the example of Kasapreko Bottling Plant in Accra, which relied on rooftop solar panels for its electricity needs (Takouleu 2019). Other institutions I visited, such as the Australian High Commission and CAL Bank, also opted for solar energy, training their facilities staff to operate and maintain rooftop panels. The Australian High Commission was the first embassy in Accra to switch to rooftop solar, and the first Australian High Commission in the world to take on this task. The facilities manager at the Australian High Commission informed me that since the installation of their solar panels, they only purchased 53 percent of their electricity from the ECG. "You stop buying your electricity from ECG, and suddenly ECG has lost a good, paying customer," a foreign aid worker who specialized in energy issues, and who had experience in the electricity sector in South America and South Asia, put these conditions in perspective for me: "The energy sector here can sink the whole economy."

As a consequence of losing more and more of its steady and paying clients and being trapped in take-or-pay contracts, Electricity Company of Ghana was forced to increase rates even more. "Imagine these high paying institutional and industrial facilities as the business class passengers on a plane," Belinda from the Energy Commission remarked, "they subsidize flights for economy class passengers, and in turn receive special treatment, say, they're offered champagne. Industrial and institutional consumers in Ghana subsidize residential consumers, and complain that they are not receiving any special treatment. We give them no champagne, so they slowly reduce their reliance on the grid." As high-paying industrial and institutional facilities invested in rooftop solar panels and minimized their reliance on national distributors, electricity consumption became a more and more demanding financial liability for participants to the grid across the country.

In this context, energy professionals aspired to expand the class of consumers who could afford rooftop solar panels. Recent scholarship on green infrastructure had singled out Ghana as a country that needed backup for its financing schemes. In a 2020 book titled *Green Building in Developing Countries: Policy, Strategy and Technology*, financing specialists wrote:

> Ghanaian practitioners have a difficult time trying to find financing sources for green projects that can defray the high costs of GBTs (green building technologies). Again, the lack of financing schemes also makes it hard to deal with the cost barrier in the GBTs adoption in Ghana. Bank loans for example are one of the most common financing schemes for green projects from around the world. Yet, within Ghana it is arduous to find banks and other financial institutions that grant loans for green projects. (Darko et al. 2020: 226)

Indicative of a particular zeitgeist, such critique has prompted the flow of relatively inexpensive financing from international organizations such as the IFC to Ghanaian banks (see IFC 2017 for a review of their green financing programs). CAL Bank, one of the large financial institutions in Ghana, started their green loan programs in 2017 with such support from the IFC.

Green finance experts at CAL Bank explained how their loan programs have not yet had this desired effect of expanding the market for rooftop solar but looked forward to future transformations. "People . . . are interested in our loans because of a lack of energy, not because they are concerned with a switch to renewables," Richard, a member of what the bankers have come to call "the green team" at CAL Bank, told me: "*Dumsor* was a great opportunity for initiating a switch to green energy, but most of us just switched to generators." Urgent solutions during *dumsor* had resulted in a transition to fossil fuel–driven electricity production across the country. Eric, another member of the green team, followed up on this argument, explaining how they searched for individuals who would establish themselves as credit-worthy, "salaried workers, who won't have problems with collateral," he specified, "but we don't have a lot of such people approaching us." Still, they aspired to develop a definition of what counted as green in Ghana and hoped to reclassify existing projects as "green projects" that would satisfy their additionality requirements.[7] "Tech for green energy is moving very fast. People are learning that energy from the national grid is more expensive [than] going solar," Eric continued, "upfront capital is high for a lot of people, but now even the presidential palace is going green."

Green finance professionals viewed the increasing palpability of climate change in Ghana as an opportunity for advancing their green loan programs. "Look at Australia," Richard offered an example, referring to the wildfire raging in the country's east coast at the time of our conversation in January 2020. Some weeks before our meeting at CAL Bank, I had participated in a panel on climate change at Accra's Jamestown Café, where the debate had concentrated on these wildfires. My fellow panelists had proposed that the situation in Ghana was not so different from Australia, as both countries had been stuck between a rock and a hard place, facing increasing temperatures and unavoidable sea-level rise. In both places, work had to be done to strengthen coastal defenses, change building codes, and determine the populations that would eventually need to be relocated. Providing a more specific perspective on Ghana, Jonathan, a climate change specialist at the Environmental Protection Agency, who often attended the climate summits as a negotiator, suggested that the country had struggled with dramatically low water levels in its hydroelectric power facilities between the years 2006 and 2010. "Water levels were so low that we were only able to keep one turbine active in the Akosombo Dam in 2008," Jonathan specified. "Temperatures are increasing, there is no water in the dams, and we are afraid of sea level rise, we need to do something about all this," members of the green team at CAL Bank agreed.

Yet such understandings of climate change issues were relatively rare in Ghana. At a time of energy scarcity, climate change issues had taken the back seat. In fact, energy professionals in Ghana repeatedly told me that they did not think they had to be held accountable for climate change impacts, as they had not participated in the more than 150 years of industrial activity that led to such high concentrations of greenhouse gases in the atmosphere. Their perspectives often followed the arguments propagated by the Kyoto Protocol, an international agreement which acknowledged that developed countries were liable for the high levels of emissions and placed a heavier responsibility on these nations in mitigating climate change.

Many of the energy professionals I met between 2016 and 2020 argued that it would be impossible to industrialize by merely relying on renewable energy systems. "How would we run an aluminum plant on solar power," many of them wondered. Ghana's iconic Akosombo Dam had been opened in 1966 with the primary purpose of providing electricity to the country's aluminum industry, but none of the contemporary renewable energy power stations could match the dam's production capacity. According to these energy professionals, industrialized countries, such as Germany, now had to rely on renewables and allow nonindustrialized nations like Ghana to build factories and other industrial facilities that required high concentrations of energy and therefore emitted greenhouse gases. "Our emissions are already so low," an electrical engineer at the Electricity Company of Ghana explained, "why should our country pay the price for the climate emergency?" While many acknowledged that Ghana already experienced the impact of climate change, such as low water levels in hydroelectric plants, they did not all agree that they had to be held accountable for the effects they experienced.

It remains to be seen how Ghana's green financing programs will transform in the upcoming years, and if they will achieve the kind of popularity that they seek to achieve among bankable consumers who might be attracted to lower interest loans. Regardless of their future, it is remarkable to observe how business models emerge as silver bullets to a variety of problems with very different causes, trying to attend to inequality, fossil fuel consumption, and climate change, all at once.

Conclusion

It is clear that most of the growth in world's energy supply and demand will take place in the global South, especially in sub-Saharan Africa. Increasing urban populations and the continent's overall economic expansion have contributed to this growth. What kinds of social and financial impact does such growing demand and supply generate? As this article has shown, the energy portfolio of Ghana transformed drastically in the past twenty years, as the country experienced low water levels in its hydroelectric power plants and faced an upsurge in electricity demands. Despite the prevalent discourse on leapfrogging to renewable energy in the global South, these emergent electricity demands have in fact incentivized Ghanaian decision-makers to enter into contracts with various independent power producers who rely on natural gas and heavy fuel oil for their electricity production. Such growth in natural gas and heavy fuel oil consumption does not denote homogeneity in Ghana's transforming energy portfolio, however. During this period of ad hoc modifications in electricity generation, some institutional, industrial, and residential consumers chose to clad their buildings in rooftop solar panels, commencing small-scale solar energy production for their own use.

In foregrounding the significance of such small-scale renewable energy production, this article specifically analyzes these rooftop solar panels. It argues that an increasing volume of rooftop solar panels installed by consumers Belinda labeled "business class passengers," that is the high paying clients of the Electricity Company of Ghana, has led to declining participation in the electricity grid, and thereby increased electricity rates for everyone else who has no choice but to remain on the grid. In response to such growing inequality, decision-makers searched for innovative business models, appealing to green loans as ways of expanding this "business class" of consumers. As a result, while a select few have managed to leapfrog to renewables, others continue to endure the grid, struggling with unsteady electricity provision and higher and higher rates.

Notes

1   This article draws on fieldwork among energy experts in Ghana between 2016 and 2020. As indicated throughout, sources for quotations include personal conversations, social media interactions, informal interviews, and public remarks. All names used in the article are pseudonyms.

2   Interestingly, scholars have not developed complex analyses of new power stations in contexts where energy is consistently scarce and difficult to access (for exceptions see Akrich 1994; Cross 2013; Mains 2012; Winther 2008). For instance, while theories of leapfrogging to renewables have been popular in imagining future energy regimes,

scholars have not researched how exactly energy related transformations occur in places where such leaps are expected to arise. One valuable strand of work focuses on the humanitarian use of renewable energy infrastructures, such as solar lamps and photo-electric lighting kits, and begins to unpack the roles they play as the material and symbolic inklings of a future leap. Although these devices are not always adopted by communities (see Akrich 1994), and therefore might not reach their end goals, according to Jamie Cross (2013: 369) they nevertheless "successfully assemble a wide array of concerns and interests, politics, moralities and ethics" (see also Dean 2020; Phillips 2020)

3   Alongside these small building-scale transformations, Chinese investors have built photovoltaic solar power stations. As of January 2020, there were two twenty-megawatt photovoltaic solar power plants in Winneba, a town about two hours west of Accra, operated by the Chinese companies BXC and Meinergy, and attached to the national grid. A third one hundred–megawatt plant was under construction in Bui, again operated by Meinergy. Overall, Chinese infrastructure projects in Ghana made up most of Chinese investment in the country, promising to create jobs for unemployed or underemployed Ghanaians and offering opportunities for low-cost technology transfer. The solar power plants promised to follow these trends. While they had very little online presence and put almost no effort into marketing, they were the largest solar power stations in West Africa. I will analyze these plants elsewhere.

4   In an article entitled "Lower electricity prices and greenhouse gas emissions due to rooftop solar: empirical results for Massachusetts" in *Energy Policy*, Kaufmann and Vaid (2016) argue that the installation of rooftop solar panels by some has eventually led to a decrease in prices among all electricity consumers in Massachusetts. A comparative analysis might allow for a better investigation as to why and how rooftop solar power does not have this impact in the Ghanaian context.

5   According to recent research by USAID's (2018) Power Africa initiative, current access rate to the grid is about 83 percent in Ghana. There is a stark distinction between rural and urban communities, where rural access to the grid is at 50 percent, in contrast to urban access at 91 percent.

6   At the time of construction, this price amounted to about $35,000. The per capita GDP of Ghana is about $750.

7   As anthropologist Aneil Tripaty (2017: 242) writes in regards to the green bonds market, such financial institutions "need to reify nature objectively to appease the bureaucratic logic of institutional investors and finance at large."

References

Ahlijah, Lom Nuku. 2017. "Can Dumsor Be Fixed? An Assessment of the Legal and Policy Framework for Privately-Owned Power Generation Projects in Ghana." LLM Long Paper. Cambridge, MA: Harvard Law School.

Akrich, Madeleine. 1994. "The de-scription of technical objects." In *Shaping Technology/ Building Society: Studies in Sociotechnical Change*, edited by W. E. Bijker and J. Law, 205–24. Cambridge, MA: MIT Press.

Awopone, Albert K., Ahmed F. Zobaa, and Walter Banuenumah. 2017. "Assessment of optimal pathways for power generation system in Ghana," *Cogent Engineering*, 4:1, DOI: 10.1080/23311916.2017.1314065

Boyer, Dominic. 2019. *Energopolitics*. Durham, NC: Duke University Press.

Cross, Jamie. 2013. "The 100th Object: Solar Lighting Technology and Humanitarian Goods" *Journal of Material Culture* 18, no. 4: 367–87.

Cross, Jamie. 2019. "The solar good: energy ethics in poor markets." *Journal of the Royal Anthropological Institute* 25, S1: 47–66.

Darko, Amos, Albert Ping Chuen Chan, De-Graft Owusu-Manu, Zhonghua Gou, and Jeff Chap-Fu Man. 2020. "Adoption of Green Building Technologies in Ghana." In *Green Building in Developing Countries: Policy, Strategy and Technology*, edited by Zhonghua Gou, 217–35. Cham, Switzerland: Springer.

Dean, Erin. 2020. "Uneasy Entanglements." *Cambridge Journal of Anthropology* 38, no. 2: 53–70.

Durodola, Abiola. 2018. "Meet Salma Okonkwo, the woman who is building Ghana's biggest solar farm." *Ventures Africa*, August 10. http://venturesafrica.com/233522-2/.

Günel, Gökçe. 2019. *Spaceship in the Desert: Energy, Climate Change, and Urban Design in Abu Dhabi*. Durham, NC: Duke University Press.

Gupta, Akhil. 2015. "An Anthropology of Electricity from the Global South." *Cultural Anthropology* 30, no. 4: 555–68.

Howe, Cymene. 2019. *Ecologics*. Durham, NC: Duke University Press.

IEA. 2019. *Seven Women Entrepreneurs of Solar Energy*. Paris: IEA. iea.org/reports/seven-women-entrepreneurs-of-solar-energy.

IFC. 2017. *Green Finance: A Bottom-up Approach to Track Existing Flows*. Washington, DC: International Finance Corporation. ifc.org/wps/wcm/connect/12ebe660-9cad-4946 -825f-66ce1e0ce147/IFC_Green+Finance+-+A+Bottom-up+Approach+to+Track +Existing+Flows+2017.pdf?MOD=AJPERES&CVID=lKMn.-t.

Kaufmann, Robert, and Devina Vaid. 2016. "Lower electricity prices and greenhouse gas emissions due to rooftop solar: empirical results for Massachusetts." *Energy Policy* 93: 345–52.

Kiunguyu, Kylie. 2018. "Ghana: Energy Maverick Salma Okonkwo Is Set to Build Ghana's Largest Solar Farm." *AllAfrica*, August 21. allafrica.com/stories/201808210308.html.

Mains, Daniel. 2012. "Blackouts and Progress: Privatization, Infrastructure, and a Developmentalist State in Jimma." *Cultural Anthropology* 27, no. 1: 3–27.

Phillips, Kristin D. 2020. "Prelude to a Grid." *Cambridge Journal of Anthropology* 38, no. 2: 71–87.

Silver, Jonathan. 2015. "Disrupted infrastructures: An urban political ecology of interrupted electricity in Accra." *International Journal of Urban and Regional Research* 39, no. 5: 984–1003.

Sorvino, Chloe. 2018. "Why This 48-Year-Old Woman Is Building Ghana's Biggest Solar Farm." *Forbes*, July 31. forbes.com/sites/chloesorvino/2018/07/31/ghana-solar-farm-ubi -salma-okonkwo/#4ca6d3dfd24.

Szeman, Imre, and Darin Barney. 2021. "Introduction: From Solar to Solarity." *South Atlantic Quarterly* 120, no. 1.

Takouleu, Jean Marie. 2019. "GHANA: CrossBoundary commissions solar off grid for bottling plant." *Afrik21*, February 16. afrik21.africa/en/ghana-crossboundary-commissions-solar-off-grid-for-bottling-plant/.

Tripaty, Aneil. 2017. "Translating to Risk: The Legibility of Climate Change and Nature in the Green Bond Market." *Economic Anthropology* 4: 239–50.

United Nations. 2015. "Paris Agreement, United Nations Framework Convention on Climate Change." December 5. sustainabledevelopment.un.org/frameworks/parisagreement.

USAID. 2018. "Ghana." usaid.gov/sites/default/files/documents/1860/Ghana_-_November _2018_Country_Fact_Sheet.pdf.

Winther, Tanja. 2008. *The Impact of Electricity: Development, Desires, and Dilemmas.* Oxford: Berg.

World Bank. 2015. "What is the Sankofa Gas Project?" *World Bank,* July 30. worldbank.org /en/country/ghana/brief/what-is-the-sankofa-gas-project.

Joel Auerbach

The Abstract Grid of Distribution:
Solar Economy beyond the Fuel Question

Long before solar irradiance could be harnessed directly to generate it, electricity stood as a Promethean challenge to the sun. As David Nye (1992: 150) relates, one of the earliest imagined promises of this technology was the abolition of night: at the dawn of electrification in the late nineteenth-century United States, the nocturnal disappearance of the sun came to be seen by some as a "check on human freedom" and productivity. Constant light would be a gift to agriculture, leading to "corn large enough to harvest with saws." Irreversibly redistributing energies, capacities, and expectations in time and space, electrification was to provide a new and uncanny foundation for life: "an illusory landscape" (3), "an impossible middle realm between nature and culture" (390). Insulated from the most basic earthly rhythms, untethered from historical toil and limitation, this new and improved sun would deliver endless abundance and liberation to those living in its relentless glow.

From the first, then, electrical infrastructures were in part about filling gaps left by the sun. In an irony of history, these same gaps now return to puncture the smooth, uninterrupted supply of electricity, at a moment when the social metabolism has grown to consume so many ter-

The South Atlantic Quarterly 120:1, January 2021
DOI 10.1215/00382876-8795815 © 2021 Duke University Press

restrial resources that perhaps only the sun itself can power it into the future. The necessity of immediate transition from fossil fuels has made painfully conspicuous the fact that energy sources such as solar and wind do not have the same material properties as coal and oil, and has attracted critical attention to the ways in which the social values and political-economic possibilities of the last century and a half have depended on a fossil regime. This makes it natural to ask what might change or become possible, in the world and in ourselves, with a mass substitution of what are often called "flow" sources for our familiar "stock" ones. While the former are inexhaustible yet intermittent and subject to fluctuations undomesticated by human will, the latter are easily broken up, shipped anywhere, and burned on command—composing what Andreas Malm (2016: 38–42) calls an "abstract spacetime," since the rhythm of consumption is independent of other natural processes. These material differences have sometimes been thought to track decisively with political-economic ones (see Scheer 2002; Schwartzman 1996). As Malm (440) has it:

> The spatiotemporal profile of the flow does not allow for anything as lucrative as the primitive accumulation of fossil capital: since the fuel is not hidden away in a separate chamber, but rather hangs like a fruit for anyone to pick, there is little surplus-value to extract in its production—no gap between the location of the energy source and that of the consumers in which the chasm between capital and labour could be reproduced.

For him, the flow represents an unequivocal "communist tendency" (379) with the potential to spur the withering away of capitalist property relations and the meeting of human needs that Marx envisioned.

   I begin this story with electricity, however, because it highlights dynamics we might miss by centering the material properties of fuel. Difficult to store in large quantities, and therefore a paragon of just-in-time production, electricity itself exists only as flow. This matters, but only as this flow is taken up, elaborated, and transformed throughout the layers of mediation composed by the electrical grid. The latter is nothing if not a massive machine for aggregating and abstracting the particular qualities and variations in energy sources, human needs, and markets, attempting to transduce these into the form of a static and eternal stock, even as the fuels that gave rise to this profile disappear. It is from within this historical, material, technical matrix that any proposed return to a life premised on solar flux will emerge and must be evaluated. One of the oft-repeated promises of solar energy is that it can eliminate long chains of extraction, profit, and pollution by effectively bypassing much of the life cycle of fuel altogether. Taking this claim hyperbolically, I want to consider solarity as a general condition that might suggest an attunement different from that of the petro-industrial age, and beyond a petro-analysis. This is not to suggest abandoning the rich analyses that attention to fuel has yielded; we are far from clear of the fossil era, and in any case transition will almost certainly mandate that solar is only one among many energies of the future. It is rather to ask, in a speculative spirit, what other questions and dynamics such a transition might bring to the fore.

This essay offers and explores three signposts for a solar analysis:

(1) Especially insofar as any question of sustainability is at stake, conditions of solarity essentially raise problems of distribution—and especially the distribution of surplus or excess. As Georges Bataille ([1949] 1991) describes, our relation with the sun inverts any economic principle that would begin from an assumption of scarcity: while lack may appear on the level of individuals, from the perspective of the total movement of energy through the biosphere, there is not just enough but too much. Solar radiation creates an inexorable pressure, an asymmetrical "gift without return," and the primary problem of such a solar or "general" economy is how to distribute and consume this excess. Although Bataille privileges consumption in his account, this only indicates a degree to which he retains a focus on the individual and wavers on the promise of general economy (cf. Groys 2012). Surpluses can be channeled and distributed in ways that either block or make space for life, with potential consequences from global poverty to war; it is the nature of this distribution that determines the reproduction and fecundity of ecosystems and that best describes the level of general movement.

(2) The relevant sense of "distribution" is abstract, functional, virtual: although materially produced (and determined by production), it refers not to matter but to the *manner* in which disparate matters hang together. It is a diagram that produces form. It is what Marx calls the "distribution of factors of production" as opposed to the distribution of products as means of consumption (Marx [1858] 2005: 94–8; [1875] 1989). Products can of course also be factors of production, and the distribution of excess, surplus, waste, and byproducts plays a particular role in reproducing the conditions of production. While this sense of distribution often depends on distribution in physical space (where it is closely tied with circulation and exchange), it refers also to the laws/norms that determine the share or apportionment of products to be assigned to different positions in *social* space (Marx [1858] 2005: 96). But only as this is itself productive or reproductive over time.

(3) While this conception of distribution is of Marxian provenance, it is as taken up in Deleuze and Guattari's *Anti-Oedipus* that I find it most useful

here. This is in part because it allows us to rethink the nature and status of subjectivity in relation to questions of energy, infrastructure, and capitalism. Thinking with this text today also invites a rigorous reassessment of the relations between concepts of flow, subjectivity, abundance, and forms of social organization. As Deleuze and Guattari ([1972] 2009: 229) describe, capital thrives on the segmentation and recombination of flows, producing a distribution of lack amidst generalized overabundance: "Flows, who doesn't desire flows, and relationships between flows, and breaks in flows?—all of which capitalism was able to mobilize and break under these hitherto unknown conditions of money." If this is true, under what conditions should we expect capital to be fazed by Malm's (2016: 367) proposed "return to the flow"?

≡

Consider the recent fate of solar energy in California. Despite famously persistent blackouts, the state actually faces an excess of electricity, spurred on by rapid development of solar energy in particular: utility-scale solar went from making up less than 0.5 percent of generation in 2010 to more than 14 percent in 2019, with rooftop panels adding approximately another 4 percent over a similar period (California Energy Commission 2020; Penn 2017). Ambitious clean energy initiatives launched in 2002 have been accelerated every few years as targets have been repeatedly surpassed ahead of schedule. Yet even as aggregate electrical demand has recently stagnated, the state government continues to subsidize construction of new fossil fuel plants at a remarkable pace. Going well beyond meeting new and expected demand, this capacity is knowingly redundant, aimed at shoring up the reliability of the grid and ensuring that there are no gaps in the available supply of electricity. Some redundancy is standard for the industry, but California's unusually large surplus is the lovechild of Clinton-era deregulatory mania and the liberal introduction of solar energy itself. In 1998, California opened its electrical generation to market competition in an attempt to decentralize its energy economy and break the monopoly power of its three major investor-owned utilities. This opened the door to increased financial speculation and market manipulation by energy trading corporations such as Enron, resulting in massive shortages and rolling blackouts in 2000–2001. Under pressure to avoid any possible repeat of this scenario, but also facing industry pressure to stay the course on deregulation, authorities have navigated this impasse by investing heavily in new generation capacity, subsidizing lucrative long-term deals for fossil fuel developers at public expense (Penn and Menezes 2017).

Pumping huge amounts of solar into this mix often simply means that solar generation is shut down first in times of excess production, because it is cheaper and quicker to turn on and off than fossil fuel plants. In other words, California burns fossil fuels that could already be replaced by existing solar capacity, even when the sun is shining, and is likely to continue to do so well into the future. During the middle of the day, the excess is often so great that the price becomes negative and California is forced to *pay* neighboring states like Arizona to take its power to avoid overloading the grid (Penn 2017). The millions of tax dollars lost to these payments, combined with the billions spent on fossil fuel plant subsidies, substantially mask the public economic benefits of solar and provide both economic and rhetorical ground for entrenched energy powers to barricade themselves from mounting sea changes. Lack is wrested from the jaws of abundance, with excess absorbed in increasingly complex financial mechanisms to shunt around power rather than in the meeting of needs. Far from making capitalization impossible, here the pressure of the sun has only ratcheted up the irrationality and discordance of the apparatus needed to maintain a semblance of stability.

Marx considered crises of overproduction to be the signature macroeconomic insanity of capitalism, proving a contradiction between the technical capacity for abundance and the social relations that organize it: production driven by independent and competing private capitals rather than actual collective need precludes rational planning and distribution of resources. On one hand, then, this phenomenon has nothing to do with solar in particular. Indeed, at time of writing, the barrel price of a major US oil future has just broken negative because COVID-19 has caused continued oil production to exceed both active demand and available storage capacity, meaning investors are willing to pay to avoid taking delivery of oil (Sheppard et al. 2020). On the other hand, a specifically solar capitalism may be one in which breakdowns of this type are routinized, completely incorporated into the everyday functioning (not just the periodic exceptions) of the machine, though still stabilized by the wicked combination of finance and statecraft that Marx's industrial crises summoned forth. It bears noting that Marx insisted on the term *overproduction* explicitly to counter the reformist explanation of underconsumption, in which the problem is simply demand collapse due to wages reflecting too little a share in the social product (Bataille may appear here on the side of the reformists). There is indeed a plausible explanation of California's impasse that places the blame on patterns of consumption: it is our unwillingness to tolerate any intermittency in the electricity supply that makes our form of life incommensurable with the rhythms of the sun. This is no doubt part of the story, but the question is how exactly subjective investments in a form of consumption function as one element in an assemblage. If the crises and absurdities in question involve a rift opening up between production and consumption, it is distribution that coordinates this disjunction.

Malm treads into this territory when he asks why fossil fuel consumption remains so firmly entrenched in our lives despite many of us knowing better. Why do more not rebel and cast it off? He suggests an Althusserian reading such that practices of fuel consumption interpellate subjects who, in being constituted by such practices, risk losing their very being in giving them up. He then concludes that consumptive practices (and therefore, perhaps—though for us this will be the question—subjectivity *tout court*), are a dead end for climate politics, because a politics addressed to individual consumer choices accords potential for change "in direct proportion to purchasing power," and by definition the wealthiest members of a fossil-capitalist society are "the subjects most thoroughly constituted by fossil use-values and therefore resistant to climate change mitigation" (Malm 2016: 362–6). But this conclusion only follows from questioning the "content" of subjectivity while taking its form for granted. What conditions ensure that consumption and subjectivity are constantly refounded on the form of the individual, market-participating subject? And indeed, on the level of content, is it really the *fuel* that subjects are invested in? This seems to ignore the infrastructural layers of mediation, alienation, and disavowal involved in such consumptive practices.

Anti-Oedipus offers a critical intervention here. Deleuze and Guattari break from the Althusserian scheme by positing a host of presubjective desires and drives: something must already exist to be called to order in interpellation. This leads them to reject the notion that desire is a matter of ideology, or of what goes on at the level of already-individuated subjective consciousness, rather, for them, "desire is a part of the infrastructure" (104). With this schema, they short-circuit the distance between the Freudian unconscious and the Marxian mode of production: though the two are not everywhere actually isomorphic, they are internally structured by the same types of syntheses/relations, and sutured together at multiple points. A key result: desire invests a social field directly, having no need to pass through certain privileged objects. "Social field" here has both a loose sense, denoting simply more-than-psychic reality (i.e., material/historical conditions), and a technical sense, which refers to distribution and to which they also apply the term *socius*. That is, desire directly invests a form of social organization and distribution, and this is what finally accounts for the productive aspect of distribution, the moment when it appears as an abstract motor that governs every force and movement. For example, capital forms a socius when, despite being produced by labor, it performs its infamous bewitchment to appear as the independent cause of all social life. A socius is thus a kind of virtual "body" of society, over which its parts/organs, flows, and surpluses are distributed:

> This is the body that Marx is referring to when he says that it is not the product of labor, but rather appears as its natural or divine presupposition. . . . It falls back on all production, constituting a surface over which the forces and agents of production are distributed, thereby appropriating for itself all surplus production and arrogating to itself both the whole and the parts of the process, which now seem to emanate from it as a quasi cause. (Deleuze and Guattari [1972] 2009: 10)

What I want to suggest is that the electrical grid, no doubt a material system (and, in places like California, deeply configured by the demands of capital), also produces an abstract body of this type, an electrified socius. We encounter it directly—not as a symptom or representation of which fuel would be an underlying reality—yet presubjectively. As constituted subjects with conscious interests, it is only under exceptional conditions (of novelty, fetish, or disappearance) that we really love electricity directly, rather than through a thousand tiny machines of consumption plugged into and detached from this massive, humming body. To say that there is also a presubjective enjoyment of this body as a whole means that we are invested directly in a disjunction, in a becoming-incommensurable with the rhythms of sun and earth: the abolition of night as a deterritorialization in the most literal sense.

It is thus not an exaggeration to say that electricity composes a *metabolic rift* (cf. Foster 2000), holding the social energy metabolism at a particular degree of intensity, and locked into an unsleeping pattern of production and consumption demanded by capital.[1] Such energetic intensity is likewise sustained by a certain pattern and intensity of desire, maintaining itself in perpetual, if barely perceptible, arousal. To use another language, the grid also produces a *plateau*: "a continuous, self-vibrating region of intensities whose development avoids any orientation toward a culmination point or external end" (Deleuze and Guattari [1980] 1987: 22); an intensity out of phase with the sun, locking the differentiated series of desire, capital, and earth into a definite form (or death spiral). The particularities of fossil fuels may have been a determining origin, but it is a serious question for energy transitions whether the path-dependent intensities and rhythms of these forms can be overcome, even if energy sources are free-flowing and abundant.

Where does this leave us? It would be unsatisfying to conclude that we simply need to desire differently, having shifted the location and nature of this desire but in the end seeking only to give it a different object—perhaps intermittency rather than constancy—as a consequence of needing to desire a different fuel. Whether or not this is viable, we should perhaps not over- state its necessity: the existence and function of the grid already confirm Malm's (2016: 376) concession that "there might be methods to engineer a more abstract profile of the flow," and perhaps this is in itself not ecologically or socially catastrophic. Proposals to invent alternative ways of inhabiting the condition of intermittency are equally worth considering. Certainly, I am with Karen Pinkus (2017) when she turns to the history of *autonomia* for tac- tics of resistance to the incessant temporality of production born in the fac- tory. A basic Bataillean proposition, for example, might be to move the bulk of industrial activity into a midday potlatch when the solar excess is at its peak. These suggestions point to the massive changes that may be necessary in the collective planning and distribution of time. There is much fertile ground here for expanding the link between energy and labor movements: recall that for Marx, a central site of struggle against capital, and the birth- place of consciousness of capitalist theft, was the cycle of the working day.

But what can the kind of solar analysis I have suggested here tell us more generally about energy, infrastructure, and capital today? *Anti-Oedipus* may seem a strange place to turn to think about form and organization, because it is so often reduced to its association with May '68 and the celebra- tion of free, spontaneous flows. Much has changed since 1968, and reading this work today perhaps requires that we place greater emphasis on the affir- mative call to build a new socius, a new body of the earth. One significant development is a continuing deterioration of faith in most of the nineteenth- and twentieth-century models for revolution on the level of molar organiza- tion (including numerous Marxist interpretations and strategies), toward which Deleuze and Guattari could position themselves as a necessary molec- ular complement. This just means we must rethink both of these aspects simultaneously today. Among much that is still vital in *Anti-Oedipus* is its reminder to find the point at which all of this touches desire and subjectivity, to discover what in desire is elaborated at the level of social production; this may be all the more urgent within an increasingly technical composition of capital that progressively excludes (in appearance, at least) any meaningful role for subjectivity.

This leads naturally to the question of why there has been so much recent critical attention to the theme of infrastructure. Dominic Boyer (2018:

224) insightfully points out that one candidate explanation is resurgent nostalgia for the Keynesian public works and welfare that buoyed many western countries from the mid-1930s to mid-1970s—an attempt at a kind of "conceptual New Deal for the human sciences." This resurgence makes sense, he observes, given its dating roughly from the 2008 financial crisis that shattered any illusion of stability for the post-Keynesian neoliberal period, which has left public infrastructures tattered and starved. Yet he is also skeptical of this nostalgia: following Timothy Mitchell, he argues that the Keynesian paradigm was tethered to a presupposition of growth that itself depended on cheap, apparently infinite oil, and "imperial control over the Middle East's oil resources . . . [by] the Anglo-American world and its allies" (224). Published in 1972, *Anti-Oedipus* was also born of (the tail end of) this era, and some have argued that it was precisely the abundance of this period that carried the May '68 movement toward questions of leisure and subjectivity as against earlier Marxist emphases on "problems of scarcity" and sustenance (cf. Bookchin [1971] 1986). But as Boyer notes, the material conditions that facilitated the Keynesian paradigm are unlikely to return. He therefore calls upon us to rediscover the category of infrastructure as revealing revolutionary potentials wholly of the present: what else could infrastructure be or portend?

Boyer then suggests that we view infrastructure as a "potential energy-storage system, as a means for gathering and holding productive powers in technological suspension" (228), basing this on Marx's description of past labor gone into a commodity as *Gallert*—an abstract, undifferentiated, and solidified mass. Yet it is curious to note how Boyer's metaphor is reminiscent of a fuel imaginary insofar as it suggests the substantialization of energy (as revolutionary potential) within a discrete object. For Marx, the revolutionary significance of machinery turned rather on the *incommensurability* of use and exchange value as two different systems of measurement: as automation requires progressively less labor, a wage (or any system that makes socially necessary labor time the criterion of distribution) becomes a more and more absurd way of linking production and consumption. It is not the accumulation of a substance, but this tension between riven incommensurables, held in an unstable distribution, that Marx viewed as potentially explosive.[2] Similarly, an emphasis on past labor is somewhat misplaced: in order to produce surplus value, capital always requires an input of current labor, however diminishingly small and obfuscated this role becomes. In fact, it is within this condition of proliferating fixed capital that abstract, molecular, transversal forms of subjectivity become especially relevant. Boyer quotes a famous statement by Lenin in which the latter says that without electricity, socialism will remain only a decreed link between the peasants and industrial working class. Lenin's implication is that the grid is exactly this link on a technical/ economic level, precisely through its capacity to abstract and redistribute across these differences. (We should recall in this connection that efforts like the 1936 U.S. Rural Electrification Act have been some of the greatest historical redistributions to counteract the intensifying division between town and country.)

Thus, my emphasis on distribution also leads back to and renews problems of production along with revolutionary class and consciousness, which any Marx-inflected political analysis must broach eventually. Ultimately, the point is to examine carefully how moments of production, distribution, and consumption are held together and apart within a constantly shifting totality, and where this creates openings for action. Such changing conditions may force us to admit that certain notions of class and "interest" no longer function in quite the old way. Indeed, part of the historical function of electricity has been to confuse, molecularize, and transversalize aspects of these operations. There is a caution here for any would-be revolution premised on the "decentralizing" and "democratizing" potentials of solar energy: if this depends on creating new solar (molar) subjects in the form of micro-proprietors invested via economic "interest" in the same socius corresponding to the market and private property, then existing incentives for overproduction, profit, speculation—and, if storage becomes more readily available, accumulation—can quite easily reterritorialize themselves on a smaller scale. This is of course not to say that a well-executed transition could not avoid this path.

A solar orientation demands starting from a perspective of Bataillean overflow, where there is always too much for anyone to hoard, and the central problem is how to channel this excess. Under what conditions could I give myself over to this dynamic without being rendered precarious by it (cf. Harney and Moten 2013: 145–6)? Could I ever trust that my needs will be met, not by the generosity of strangers, but by their own inexorable need to give something away? Certainly in California today, systems of market and private property, and the boundary-lines of state, corporate, and individual entities they depend on, are so deeply ingrained that it seems impossible, if not outright insane, to find ways to dispose of excess power as a free gift, rather than pay others to take it. This really would imply a radical redistribution of time, expectation, and the boundaries of subjects/entities. The perspective of general economy is not a unitary holism without subdivision, but a system of breaks and flows; parts are less subordinated to a whole than dissolved in their self-possession by a transversal movement they cannot contain. If there is indeed a "communist tendency" to this kind of flow, it must be thought not through a privileging of absolute movement but through a pragmatic capacity to redraw the diagram of an abstract grid of distribution, and rewrite the principles of movement across it. Solarity need not mean blowing in the breeze, or flowing any which way. There is also a disciplinary function of excess, a sober and humble experimentation required to compose with this blinding, deterritorializing force. Glaring down, its command is insistent: consent not to be one, scurry to new ground, make yourselves ready for the coming night.

Notes

1   Marx's concept of metabolic rift already turned on the economic moment of distribution in the foregoing sense. Nutrients from rural agricultural production flow into cities, where they accumulate as waste and pollution, rather than being returned to the soil to restore its capacity for future production. Lacking the means to rationally plan and redistribute these flows, capitalism degrades the potentials of the earth in much the same manner that it degrades labor-power. Greenhouse gases can be understood in much the same way, as an unintegrable free radical that deteriorates the reproductive capacity of ecosystems (when stimulated by solar heat). The distribution of waste and byproducts may thus hold a better key to understanding climate change than accounts based (often tacitly) on overproduction or overconsumption as isolated moments. A whole study could be made of this theme in *Anti-Oedipus* as well: the strange connection between shit and ideality, the state as alternately a rationalizing force or absolute death-drive, the substitution of anal productions and distributions for the phallic signifier . . .

2   This bespeaks an easy misreading of the metabolic rift school, which, in fairness, may here stem from a quotation from Foster and Paul Burkett that is vague out of context: "Of course, this value (energy) surplus is not really created out of nothing. Rather, it represents capitalism's appropriation of portions of the potential work embodied in labor power recouped from metabolic regeneration largely during non-worktime" (quoted in Boyer 2018: 228). What is unfortunate here is the suggestion of an equivalence between value and energy; rather, these should be taken to designate separate, incommensurable series. For Marx, exchange value is an accounting system for social labor time. It is not that energy (and Marx does sometimes mean literal muscular energy, on one side of the disjunction) is literally transformed into a substance called value (which is then entirely mysterious, a metaphysical entity), but that it is registered in a different manner of being, as information (or what Deleuze and Guattari call "recording" or "inscription"). It is true that capitalists get more energy (labor) than they pay for; however, as *value* it is only a representation of that work, and value and labor are differentially apportioned according to a preexisting distribution of social relations. These authors are exceptionally clear on this point and its importance elsewhere (Foster 2018; Foster and Burkett 2008).

References

Bataille, Georges. (1949) 1991. *The Accursed Share, Vol. 1: Consumption,* translated by Robert Hurley. New York: Zone Books.

Bookchin, Murray. (1971) 1986. *Post-Scarcity Anarchism.* Montréal, QC: Black Rose Books.

Boyer, Dominic. 2018. "Infrastructure, Potential Energy, Revolution." In *The Promise of Infrastructure,* edited by Nikhil Anand, Akhil Gupta, and Hannah Appel. Durham, NC: Duke University Press.

California Energy Commission. 2020. "2019 Total System Electric Generation," June 24. energy.ca.gov/data-reports/energy-almanac/california-electricity-data/2019-total-system-electric-generation.

Deleuze, Gilles, and Félix Guattari. (1972) 2009. *Anti-Oedipus: Capitalism and Schizophrenia,* translated by Robert Hurley, Mark Seem, and Helen R. Lane. New York: Penguin.

Deleuze, Gilles, and Félix Guattari. (1980) 1987. *A Thousand Plateaus,* translated by Brian Massumi. Minneapolis: University of Minnesota Press.

Foster, John Bellamy. 2000. *Marx's Ecology: Materialism and Nature.* New York: Monthly Review Press.

Foster, John Bellamy. 2018. "Marx, Value, and Nature." *Monthly Review* 70, no. 3. monthlyreview.org/2018/07/01/marx-value-and-nature/.

Foster, John Bellamy, and Paul Burkett. 2008. "Classical Marxism and the Second Law of Thermodynamics: Marx/Engels, the Heat Death of the Universe Hypothesis, and the Origins of Ecological Economics." *Organization and Environment* 21, no. 1: 3–37.

Groys, Boris. 2012. *Under Suspicion: A Phenomenology of Media,* translated by Carsten Strathausen. New York: Columbia University Press.

Harney, Stefano, and Fred Moten. 2013. *The Undercommons: Fugitive Planning and Black Study.* New York: Minor Compositions.

Malm, Andreas. 2016. *Fossil Capital: The Rise of Steam Power and the Roots of Global Warming.* Brooklyn, NY: Verso Books.

Marx, Karl. (1875) 1989. "Critique of the Gotha Programme." In *Marx/Engels Collected Works, Vol. 24.* London: Lawrence and Wishart.

Marx, Karl. (1858) 2005. *Grundrisse: Foundations of the Critique of Political Economy,* translated by Martin Nicolaus. London: Penguin Books.

Nye, David E. 1992. *Electrifying America: Social Meanings of a New Technology, 1880–1940.* Cambridge, MA: MIT Press.

Penn, Ivan. 2017. "California Invested Heavily in Solar Power. Now There's so Much That Other States Are Sometimes Paid to Take It." *Los Angeles Times,* June 22. latimes.com/projects/la-fi-electricity-solar/.

Penn, Ivan, and Ryan Menezes. 2017. "Californians Are Paying Billions for Power They Don't Need." *Los Angeles Times,* February 5. http://www.latimes.com/projects/la-fi-electricity-capacity/.

Pinkus, Karen. 2017. "Intermittent Grids." *South Atlantic Quarterly* 116, no. 2.

Scheer, Hermann. 2002. *The Solar Economy: Renewable Energy for a Sustainable Global Future,* translated by Andrew Ketley. Sterling, VA: Earthscan.

Schwartzman, David. 1996. "Solar Communism." *Science & Society* 60, no. 3: 307–31.

Sheppard, David, Myles McCormick, Anjli Raval, Derek Brower, and Hudson Lockett. 2020. "US Oil Price below Zero for First Time in History." *Financial Times,* April 20. ft.com/content/a5292644-958d-4065-92e8-ace55d766654.

India Citizenship Law

Ranabir Samaddar, Editor

Ranabir Samaddar

Revolution and Counterrevolution in India

This special section discusses the protests against the anti–Citizenship Amendment Act (CAA) in India, passed on December 11, 2019. The protests continued roughly from November 2019 to mid-March 2020. The five entries in this section tell us of a time of tumult, citizens' upsurge across the country, the nature of a state that legitimates itself by being an exception to a less than adequately nationalized society, the nature of public protests and the specific and varying nature of protest in different parts of the country, the imagination of the country's constitution as commons that enabled popular interpretations, understanding, interventions in the constitutional domain, and, finally, the postcolonial aporia of the citizen/alien duality that produces precarious lives in the entire region of South Asia. The CAA apparently was intended to clarify the confusion emerging out of this interlocked duality once and for all.

It is not even mid-June, barely three months later in the year. Yet, in the midst of the COVID-19 pandemic, the time of anti-CAA protests looks distant. Are these two phases—the phase of seething popular protests against the CAA with its groundswelling nature and the succeeding phase of total demobilization of politics during the countrywide total lockdown—disconnected, accidental in their association, with one effacing the other? Whereas the former phase was characterized by dense political participation of large sections of society, the latter phase is marked by banishment of politics, stripping the great question of life of any reference to politics, and an overwhelming centralization of the bureaucratic-political apparatus of rule over the life of the people. Although future historians will debate over the quality as well as the fault lines of the time, I have no doubt that the time

The South Atlantic Quarterly 120:1, January 2021
DOI 10.1215/00382876-8795830 © 2021 Duke University Press

from November 2019 to this day, when this small introduction is being written, constitutes a single time, known by that much abused word—*crisis*.

What is this *crisis* to which these six months gesture and to which the first three months of this six-month-long period—the months of protests—belong? At one level, it all began with the general elections of 2019 (April–May 2019) when the central government announced to the populist opposition in various parts of the country: in the event it was returned to power, it would ensure that the law of the land would prevail, illegal aliens would be expelled, unruly elements would be straightened out, the country would be made properly nationalist, and the economy would be unlocked with the tools of aggressive privatization, support to corporate power, labor deregulation, and fiscal tightening that any orthodox monetarist would have been proud of. The credit market was squeezed, indebtedness among common people rose with welfare expenditure going down and banks reducing their operations, the GDP declined rapidly, the economy shrank with each passing day, and the Left-liberal opposition—the alternative nationalist force—fell before the repeated invocations of legality, judiciary, fiscal responsibility, and the norm of responsibility. The populists entrenched in states, localities, and specific domains of political life were the only troublesome factor in the new model of power, which was based on a combination of nationalism and globalization, "made in India" strategy and open entry to global capital in all sectors of the economy, "country first" and privatization, and finally law and authoritarianism. Populists represented what has been called illegalisms of society. And, we must not forget that these populist forces hosted, sheltered, encouraged, and facilitated the protests against the CAA, leading to an upsurge of minority communities against the new model of power. We have to think deeper about the dynamics of the spread of the protests in a quick span of two months—December 2019 and January 2020.

The Delhi riots followed the popular upsurge. For a week beginning on February 23, multiple waves of bloodshed, property destruction, looting, and arson overwhelmed Delhi, for some time the epicenter of anti-CAA protest. What had happened to protesting Muslims in towns in eastern Uttar Pradesh in the past two months now openly began in Delhi. About fifty-five people were killed, two-thirds of them Muslims—shot, slashed, hammered with repeated blows, or set on fire. The dead also included a police officer, an intelligence officer, and over a dozen Hindus, all of them shot or assaulted. Hundreds of wounded languished in inadequately staffed medical facilities, and even a week later, corpses were being found in open drains. Later, many Muslims were found missing.

In one blow, the anti-CAA protest had been quelled. The political parties once more proved ineffective before the power of legality and authority. The Hindu template of an authoritarian nationalist power had proved effective. But we must not be surprised at this near-simultaneity of "revolution and counterrevolution." The historically minded would remember how a radicalized city of Kolkata, fervent with anti-colonial protests demanding the release of INA (Indian National Army) prisoners, quickly turned into a site of communal holocaust in 1946. Likewise, a radicalized year in 1974, marked by the largest railway-worker-led general strike in the country's history and the setting up of "Janata Sarkars" (people's governments/councils) in the state of Bihar, was followed by the imposition of the National Emergency in 1975. The crisis has always been resolved one way or another. History does not wait for the Left or the liberals.

Here in India, the political crisis accompanied by the fiscal-economic crisis met with the epidemiological crisis in late February–March. The impending epidemiological crisis provided the government the occasion to invoke the 1897 Epidemic Diseases Act (popularly known as the Plague Act) and the 2005 National Disaster Management Act. The template of managing COVID-19 in India by an extremely centralized governmental apparatus was set. Policies of law and order took the place of public health. The measures related to the novel coronavirus pandemic stemmed from such a response. Politics was to be suspended. Once again, the traditional political forces lowered their guards. The result was the tragic "migrant crisis," images of which beamed all over the world, showing thousands upon thousands of migrant laborers trekking hundreds of kilometers to escape hunger, joblessness, and absence of shelter with three hundred-odd workers dropping dead on the road or being run over by trains or speeding buses and trucks. Meanwhile, the public health scenario remains grim. States, localities, slums, villages, small towns, and city councils have emerged as the frontline defense. They are fighting on their own against the epidemiological specter.

Yet this is the perfect neoliberal occasion of, as is said, "never let a crisis go waste." The spate of measures announced by the government in May 2020 is perched on a single-minded drive to carry through the suspended reforms to make the country safe from pathogens and underdevelopment. And once again, labor will be the raw material on which the reforms will be experimented, planned, and executed.

At one level, it is a story of earlier histories of crisis resolved this way or that. Yet, possibly this time, the crisis, itself a combination of several crises, has raised the important question: What is the political horizon that will encourage us to imagine a new kind of society—a caring society whose mandate will be to foster care and nondiscrimination? What will be the imaginary of a new kind of power that is dialogic and takes upon itself the task of protecting the vulnerable, and thus help develop a new meaning of life? The two phases of the last six months in India's life pass into one another, forming one congealed moment of contentious politics.

Of course, we know that the title of this short note is exaggerating the situation. But recall how Friedrich Engels ([1851] 1896: 2) analyzed revolution and counterrevolution in Germany after 1848:

> If, then, we have been beaten, we have nothing else to do but to begin again from the beginning. And, fortunately, the probably very short interval of rest which is allowed us between the close of the first and the beginning of the second act of the movement, gives us time for a very necessary piece of work: the study of the causes that necessitated both the late outbreak and its defeat; causes that are not to be sought for in the accidental efforts, talents, faults, errors, or treacheries of some of the leaders, but in the general social state and conditions of existence of each of the convulsed nations.

Now, too, we need to study the causes of rebellion and defeat so that a second act of the movement may open.

References

Engels, Friedrich. (1851) 1896. *Revolution and Counter-Revolution in Germany*, edited by Eleanor Marx Aveling. London: Swan Sonnenschein and Co. marxists.org/archive/marx/works/download/pdf/revolution-counterrevolution-germany.pdf.

Partha Chatterjee

The State of Exception Goes Viral

When, in May 2019, Narendra Modi was elected to a second term as prime minister, his government signaled a change of course. His first term had begun in 2014 with a promise of economic reforms to satisfy the expectations of big business and the upper middle class. But the global slump left little room for the government to drastically change tax and labor laws or cut down social expenditure. Instead, faced with growing unrest from farmers and other regional lobbies, Modi had been forced to fall back on time-tested methods of populist distribution of benefits to large, electorally mobilized groups. Returned to power, however, with Amit Shah, president of the Bharatiya Janata Party (BJP) and acknowledged master of political dirty tricks, by his side as the new minister of home affairs, he took little time to announce that his government would now take a series of steps to realize the party's ideological dream of instituting a Hindu nation-state.

The March toward a Hindu Nation-state

The first major move was to revoke the special constitutional status of the Muslim-majority state of Jammu and Kashmir. This status, embodied in Article 370 of the Indian constitution, was part of a series of negotiated agreements by which nearly six hundred princely states, ruled by Indian rulers with whom the British had made treaties, were integrated into the Indian union after independence in 1947. There were other states that also had special status, as indeed do some even now. But Article 370 was always regarded by Hindu nationalists as an eyesore because it seemed like a bribe to the Muslims of Kashmir to induce them to stay with India, especially in view of

The South Atlantic Quarterly 120:1, January 2021
DOI 10.1215/00382876-8795842 © 2021 Duke University Press

the bitter rivalry over the territory with Pakistan. The BJP had often demanded that Jammu and Kashmir be treated like any other state of India, with no special protection for domiciles of the state. It was alleged by some that the BJP plan was to allow people from other states to settle in large numbers in Kashmir and thus change its demographic character. In August 2019, not only was Article 370 rendered irrelevant but the Buddhist-majority district of Ladakh was separated and both parts of the erstwhile state were reduced to the status of union territories, governed directly from New Delhi. There was a complete lockdown enforced by hundreds of thousands of armed troops and the entire political leadership of Kashmir was put behind bars. Although a few leaders have been conditionally released, the situation remains largely unchanged today, ten months later.

The sudden decision against Jammu and Kashmir was met with widespread jubilation in cities across northern India where large sections, still enthused by the memory of the near-war with Pakistan a few months earlier that boosted Modi's popularity before the elections, saw this move as a slap in the face of Pakistan and its supporters in Kashmir. For a long time now, the public in India, even those who were otherwise critical of the BJP, had come to regard Kashmir as a piece of real estate that legitimately belongs to India and the people of Kashmir as perpetual troublemakers. Still stunned by the unexpected scale of the BJP's election victory, the opposition was largely muted in its response. The Modi–Shah duo then went on to apply to individuals the laws against organizations allegedly engaged in unlawful activities, making it easier to put inconvenient critics in prison without trial. (The law has been recently applied to several journalists and rights activists.) An added boost was the judgment of the Supreme Court in November to allow the construction of a Hindu temple on the site where a four-hundred-year-old mosque had been destroyed by Hindu nationalists in 1992. Modi's government immediately announced plans to facilitate the building of the temple.

In the meantime, a festering problem in the northeastern state of Assam reached a crisis. Following a prolonged agitation there against the incursion into the state of non-Assamese people, most alleged to be illegal migrants from Bangladesh, an agreement had been reached in 1985 to verify the antecedents of each person living in Assam and compile a register of citizens. The exercise was hobbled by numerous legal and bureaucratic difficulties. Since there was no such thing as a national identity card and birth certificates were rare until very recently, various documents such as ration cards, voter identity cards or titles to property came to be accepted in this state with international borders as proofs of domicile amounting to naturalization into citizenship. In 2003, an earlier BJP-led government in New Delhi amended the citizenship act to prevent the conferring of citizenship on any illegal migrant who had entered into or was living in India without a valid visa. After repeated prodding by the Supreme Court, the exercise of compiling the National Register of Citizens was completed in Assam and the lists were released in September 2019.

The results came as a shock. Of the 1.9 million who were declared illegal migrants, 1.2 million were Hindus and 700,000 Muslims. This put the BJP in a fix since it could not now detain or deport such a large Hindu population, mainly from Bangladesh, since it had always claimed they were refugees who had fled from religious persecution. This was the immediate impetus for pushing the Citizenship Amendment Act (CAA) through parliament in early December 2019. The amendment laid down that non-Muslims coming to India from Afghanistan, Pakistan, and Bangladesh will not be regarded as illegal migrants and could be given citizenship through an expedited process. Amit Shah explained in parliament that the new citizenship law would be followed by the compiling of a National Register of Citizens throughout the country.

It is important to note that the conception of citizenship embraced by political leaders in the early years after independence was *jus soli*—citizenship by birth on the country's soil. The constitution declared in 1950 that everyone living in India who was not a citizen of another country was a citizen of India. The citizenship law of 1955 gave citizenship to everyone born in India. From the 1980s, however, the consensus seemed to shift towards the recognition of *jus sanguinis*—the ethnicity of one's parents—as the more powerful claim. The 2003 amendment laid down that those born after 1987 could only become citizens if at least one parent was an Indian; for those born after 2003, either both parents had to be Indian, or one Indian and the other not an illegal migrant. At the same time, the 2003 amendment conferred on people of Indian origin who were citizens of other countries (except Pakistan and Bangladesh, of course) the status of Overseas Citizens of India, which gave them most rights enjoyed by Indian citizens except the right to vote in Indian elections. This created a large pool of affluent diasporic converts to the cause of Hindu nationalism.

The Anti-CAA Protests

It was patently obvious that these moves of the BJP government were biased against Muslims. They fitted with its ideological position voiced from before independence. V. D. Savarkar, one of the BJP's nationalist heroes, writing in

1923, defined Indian nationalism as Hindutva or Hindu-ness and argued that only those who regarded India as their fatherland as well as holy land could belong to the nation. This definition excluded Muslims and Christians who, he declared, regarded other places as their holy land. M. S. Golwalkar, another ideological icon, wrote in 1939 that those who were not culturally Hindu could only live in the Hindu nation-state as domiciles without rights of citizenship. Following the partition of the country, Hindu nationalists often argued that all Hindus should be brought from Pakistan and settled in India and all Muslims sent from India to Pakistan. In recent years, they have referred to the presence of Muslims in India as the unfinished business of partition. Even though Narendra Modi was carefully kept out of such blatant anti-Muslim propaganda, other BJP leaders seldom shied away from spreading hatred against Muslims, leading to numerous murderous attacks on Muslims by BJP vigilante groups.

The passing of the Citizenship Amendment Act (CAA) in December 2019 was thus a particular cause of alarm among Muslims all over India. The fact that the granting of citizenship was being defined in terms of religion and the warning that it would be followed by a National Register of Citizens (NRC), in addition to the experience of Assam where millions of residents had been declared illegal migrants because, in the opinion of state officials, they did not have adequate documents, led to widespread panic that Muslims would now be targeted. Curiously, the panic was not restricted only to Muslims. Border states, like West Bengal, for instance, where many Hindus had moved from erstwhile East Pakistan and later Bangladesh several decades ago and were, as far as they knew, fully naturalized into Indian citizenship, were thrown into confusion, and the example of Assam was far from reassuring. No matter how hard BJP leaders tried to argue that the CAA was intended to grant citizenship, not to take it away from genuine citizens, the doubts would not go away. If there was a countrywide NRC, everyone's citizenship would be put to the test.

The protests against the CAA began sporadically in different states from the middle of December 2019. In northern India, they began from certain Muslim neighborhoods in Delhi and Uttar Pradesh. In the latter, the BJP state government cracked down viciously on the protesters. The reporting was muzzled to such an extent that the true scale of police violence is still unknown. Student protesters were also beaten up by the police on at least three major university campuses. These scenes of police brutality were widely circulated, galvanizing opinion against the CAA in cities and towns across the country.

The mobilization that took place over the next ten weeks or so was unique in many ways. It was mostly organized and peopled by those who had never participated in political rallies before; many claimed to be uninterested in and uninformed about politics. Particularly noticeable was the large presence of Muslim women of all ages who occupied street corners and parks for weeks to declare their refusal to produce documents to prove their citizenship. But they were not alone. They were supported by people of all ethnic identities, especially the young, who repeated the opening sentences of the constitution that declared that the people of India, regardless of language, religion or caste, had given themselves the republic. The fact that these protests were peaceful and involved people without any obvious political affiliation made the BJP government wary of using force.

It was striking that the opposition parties, which were initially clueless on how to respond to the CAA, slowly joined in support of this largely spontaneous agitation. But they could not agree on what to do next. A few states ruled by regional parties announced that they would not assist in the NRC exercise in their states. But they had other electoral calculations to make and seemed reluctant to be characterized as partisans in a cause in which Muslims were such a visible presence. During the latter half of January 2020, there was a high-pitched campaign in Delhi for elections to the state assembly. The ruling Aam Aadmi Party (AAP), tipped to win, trod carefully around the issue while the BJP accused it of inciting the large anti-CAA protests in the capital. AAP swept the elections in early February.

Throughout the election campaign in Delhi, BJP leaders had spewed venom at the protesters, and the so-called anti-nationals and urban militants who allegedly mobilized them, often calling upon BJP supporters to "shoot the bastards." After the BJP defeat, they demanded that the police immediately disperse the protest gatherings; otherwise, they would do so themselves. On February 24, 2020, when Donald Trump was being feted by Modi and Shah at a mammoth rally in Gujarat, the attacks began in Delhi on Muslim neighborhoods. For the first day or two, there was something of a fight between the two sides, armed with bricks, stones, and homemade firearms. In subsequent days, it was a pogrom against Muslims. Houses and shops were set on fire and people were slashed, stoned, or burnt to death, as the police literally stood by and watched. The official death toll was fifty-three, thirty-six of them Muslim.

The Delhi violence made abundantly clear the dangerous course that the BJP had adopted. It had brazened out the anti-CAA protests, not conceding an inch to calls to rethink the citizenship issue. Even as it suffered a string of electoral defeats in the states, it tightened its grip on the federal administrative machinery and redoubled its propaganda efforts through a variety of media. In mid-March, Amit Shah engineered defections from the ruling Congress Party to bring down the government in Madhya Pradesh and install a BJP chief minister. The coup must have required considerable time and effort because it delayed the official response to the COVID-19 pandemic.

Political Consequences of the Pandemic

The first COVID-19 cases were reported in late February among travelers from Europe. Within a week, cases began to be noticed in several parts of the country: most seem to have originated from those who had arrived from Europe and the Middle East. But it took until March 23 for the prime minister to appear on television to announce a countrywide lockdown. The administrative confusion that followed showed every sign of unpreparedness. The idea seemed primarily to protect the urban middle classes, both physically and economically, from the consequences of shutting down the economy. Since a significant number of middle-class people were employed, either directly or indirectly, by the government, their incomes were protected even when they were unable to report for work. Large private employers were coaxed to join the national effort by not firing their employees. Much office work shifted to various forms of "working from home." The blow was felt most severely by the huge mass of so-called informal workers, most of whom are self-employed with little or no savings, and many temporary migrants in locations that were hundreds of miles away from their homes. Since supply chains were thoroughly disrupted, those in small trade and agriculture were in dire straits. And with the winter crop ready to be harvested, there was panic in the countryside because there were not enough laborers and no assurance that the crop could be marketed.

For some time now, a massive fault line has opened up in India between the formally organized economy following the usual logic of capitalist accumulation and the informal economy based largely on subsistence through market exchange. The latter, however, is not the traditional economy, which has practically disappeared, but a creation of what Marx called the primitive accumulation of capital. The millions in the informal sector are not a necessary part of the formal economy dominated by corporate capital and are only tenuously connected to it. They survive because, by participating in organized electoral politics, they are able to secure some protection for their livelihoods from government that provides them with cheap food and other necessities and tolerates minor violations of regulations on labor, pollution, public health, or taxes. The competitive populism that characterizes Indian electoral politics is what sustains this massive population.

The unprecedented and exceptional nature of the response of the government in New Delhi to the health emergency has ripped off its populist facade. Hundreds of thousands of migrant workers left without work or housing decided to walk hundreds of miles to go home: no one knows how many died on the way. At the same time, special flights were arranged to bring back students and tourists stuck in China and Europe. The prime minister appeared on television from time to time to invite people to stand on their balconies and applaud the health workers or light lamps to scare away the monster virus: those who have balconies to their houses—the urban middle class, that is—excitedly followed their leader.

To be fair, the COVID-19 epidemic is so unprecedented that governments everywhere are fumbling in their response. What is particularly hazardous for India in the near future is the political response to the inevitable economic collapse that is looming. All politics is suspended as Modi speaks of a war on the new enemy. The attempt is to create a centralized structure of command that would conduct the war in every state, every district, and every neighborhood. As a police officer in Uttar Pradesh said in a public order, later withdrawn: "This is not the time to exercise your judgment or intelligence. This is the time to follow orders." The wheels of government are being oiled and tuned to function as though it was a state of emergency in which normal rules are suspended.

But the strains are showing as the government tries to balance the demands of health management with the dire projections of an imminent economic collapse. The visible desperation of migrant laborers finally forced the authorities to arrange for trains to send them home. But in the meantime, the pandemic, contained for a while, had begun to spread exponentially. State governments, which bear the primary responsibility of caring for the lives of people, suddenly began to police their borders and prioritize the needs of their own populations. Yet it is the central government that controls the purse strings. Modi's government seems bent on tightening executive control through the central bureaucracy while passing on to the states the responsibility of dealing with the grievances of the people. In the absence of normal politics, the state of emergency is being formalized. The Hindu nation-state might soon be realized without the paraphernalia of electoral democracy.

Bharat Bhushan

Citizens, Infiltrators, and Others:
The Nature of Protests against
the Citizenship Amendment Act

Although the Citizenship Amendment Act (CAA) offers speedy asylum and citizenship to non-Muslim minorities from India's three neighboring countries, it has made the naturalization of Muslims from those countries more difficult. The preparation of the National Population Register/ National Register of Citizens (NPR/NRC) is an equally ideological exercise designed to monitor, sequester, and threaten India's minority Muslim community with disenfranchisement. Protests against the CAA and NPR/NRC emerged spontaneously all over India. In India's northeastern states protestors sought protection for the political, cultural, and land rights of the indigenous population against all immigrants regardless of religious affiliation. In the rest of India, however, protestors questioned whether the secular nature of the Indian Constitution was undermined by basing citizenship rights on religion. Muslims feared these measures could erode their citizenship rights. The most iconic of these protests was a 101-day sit-in at Shaheen Bagh, a Muslim dominated area on the outskirts of Delhi. Led by women, it spawned similar protests across the country. While no political party took leadership of these protests, perhaps reflecting the growing weight of majoritiarianism in India's electoral politics, the popular movement itself threw up a new and youthful leadership. It fashioned a new vernacular idiom of protest, choosing secular civic symbols over religious identity-markers, and created a new civic consciousness rooted in constitutionalism.

The South Atlantic Quarterly 120:1, January 2021
DOI 10.1215/00382876-8795854   © 2021 Duke University Press

Political Motivations

We should be clear that the CAA is not a response to a steep increase in religious persecution in India's neighboring countries and that the National Population Register/ National Register of Citizens (NPR/NRC) is not a response to India being overrun by illegal migrants. These significant changes in India's citizenship laws are occasioned by quite different concerns of the ruling Bharatiya Janata Party (BJP). The CAA has especially been criticized for creating a legislation where executive policy decisions would have sufficed in granting citizenship quickly. Standard procedures for this exist and have been utilized in the past. Moreover, the law leaves out of reckoning persecuted minorities in several other neighboring countries, such as Tamils in Sri Lanka, Rohingyas in Myanmar, Lhotshampas (Bhutanese of Nepal origin) in Bhutan, and Tibetans in Tibet.

These far-reaching changes in the framework of citizenship are linked with the BJP advancing the goal of a majoritarian state for its voters and associating it with nation building. It has already fulfilled the key promises of its electoral agenda. Jammu and Kashmir has been fully "integrated" into the Indian Union. Instant triple-*talaq* (the Muslim practice of oral divorce by men uttering the word "*talaq*" thrice to abandon their wives) has been criminalized and enacted into law by Parliament. The Supreme Court of India has cleared the way for the construction of a Ram Temple at a disputed site where a mosque had been destroyed by Hindu zealots to claim the land for a Hindu temple in 1992.

The only item left on the majoritarian agenda was to implement a Uniform Civil Code, which has been a deeply divisive issue since the 1920s, as the Muslim community largely wants to follow a religiously defined personal law relating to marriage, divorce, and adoption. However, the Uniform Civil Code will be difficult to implement any time soon as it is caught up in legal and other consultative processes. The party, therefore, needed to line up a new program to rally its Hindu voters, both for the frequent state elections and, more importantly, something that would last the course till the next general election in 2024. The CAA and NRC offered such an agenda.

The CAA promising a fast track citizenship for non-Muslims may have been accidentally occasioned by the failure of the NRC in Assam where a limited exercise was mandated many years ago to detect Bangladeshi illegal migrants. To the chagrin of the BJP, the NRC failed to locate a large number of Muslims among illegal migrants in the state. However, the BJP realized the political potential of a registry if it were implemented nationally.

The CAA, in addition to making the naturalization of Muslim migrants from the neighboring countries difficult, would, at an ideological level, establish the notion of India as a "Hindu Homeland" or a Hindu nation to which the BJP is doctrinally committed. The NRC would create the occasion to speak of Muslims as a threat to the nation by highlighting their links with their co-religionists outside the nation's borders. Since the population of Muslims is diffused across the geography of India, the NPR/NRC offered a way to acquire knowledge about their location, to monitor them in the future, sequester them, and when necessary, disenfranchise them.

Two Streams of Protest

The enactment of the CAA spawned two broad categories of popular protests. The protests in India's northeast were qualitatively different from those in the rest of India. In the northeast, these protests were spearheaded by indigenous/ethnic groups who feared they would be demographically swamped by in-migration. Issues of identity define politics in these states, and so the protestors wanted no citizenship rights to be extended to any refugee/immigrant, regardless of their religion. The preservation of the political, cultural, and land rights of the indigenous population they felt would be further threatened by the CAA as it could potentially trigger fresh migration of Hindus and Buddhists from Bangladesh into the northeastern states.

The protests in the northeast were led by college and university students—the All Assam Students' Union, Naga Students' Union, Arunachal Pradesh Students' Union, Mizo Zirlai Pawal, Khasi Students' Union of Meghalaya, Twipra Students' Federation of Tripura, and North East Students' Organisation—as well as by women's groups, litterateurs, artistes, and public intellectuals. The Union government seems to have been prepared to deal with expected protests in the northeast. The CAA did not apply to Arunachal Pradesh, Mizoram, and Nagaland. Nor did it apply to the Sixth Schedule areas of Meghalaya, Tripura, and Assam protected by the Inner Line Permit (ILP) system that prohibits outsiders from purchasing land and requires visitors to get a special travel permission.

As a result of the protests, the government decided to extend the ILP system to Manipur as well. The Assam government created three new Autonomous Councils for the Koch Rajbanghis, Moran, and Muttock communities. Reservations in educational institutions and measures for securing their land rights were also announced. These concessions in the ILP and other measures in Assam to an extent dampened the fury of the anti-CAA movement in India's northeast. The Modi government, however, appeared completely unprepared for the intensity and the spread of the protests in the rest of India. These were much less inward looking than those in the northeast and the protesters' objections to the new citizenship law were on constitutional grounds. They felt that basing citizenship on religion violated the basic principle of granting citizenship up to now. The CAA also violated, they felt, Article 14 of the Constitution of India, which forbids the State from denying any citizen equality before the law and equal protection of the laws, and does not allow discrimination on the grounds of religion, race, caste, gender, or place of birth.

These protest movements also expressed dismay that the social consequences of the CAA would be divisive and destroy social amity. The Muslims expressed concerns that they were being relegated to a lesser class of citizenship by the state's preferential treatment of non-Muslims in the CAA. They also feared that the proposed exercise of a countrywide NRC might render many Muslims stateless. As the experience of NRC in Assam showed, the documents required to prove one's citizenship would push the Muslim community to the wall relative to the majority community, which could count on the enumerator's discretion to be included as citizens in the NRC.

The government could only give weak assurances to the protestors that it had no immediate plans for preparing a nationwide NRC. The critics of the government pointed out that the NPR exercise had added six new questions to the questionnaire first used in 2010. These questions sought details of parents' birthplace, mobile phone number, Voter Identity number, mother tongue, and Aadhar number—a twelve-digit number issued to the residents of India (by the Unique Identification Authority of India), which gives access to biometric details of the person. In response to protests and apprehensions about the NPR in the minds of the minority community, many state governments said that they would not implement the NRC and that only the old NPR (2010) format would be used for enumeration.

The sites of the protests against citizenship laws (both CAA and NPR/NRC) were largely universities and metropolitan cities. The spark that lit a wildfire of protests was the brutal beating of students by Delhi Police, which had forced its entry in the library of Jamia Millia Islamia University. The video captures of police brutality on hapless students soon spread across the country. The newspaper photograph of Muslim girls wearing hijab shielding a male student by taking on a baton-charging police officer became an iconic symbol of the protests.

In the universities, the protests were led by students' unions and by academics. In other metropolitan city spaces, the protests were led by public intellectuals and civil liberties activists. They were joined by amorphous groups of youngsters who were participating in public demonstrations against the government for the first time in their life.

These protests became a celebration of India's inclusiveness and secular character with a festival-like atmosphere marked by impromptu music concerts, singing, street theatre, and, iconically, by public reading of the Preamble of the Constitution of India. Youngsters turned up with homemade, hand-painted posters and poems and songs that condemned the government's divisive politics. A memorable poster, later to reappear in many protests and social media posts all over India, read, *Jab Hindu Muslim razi to kya karega Nazi* (When Hindus and Muslims are one, what can the Nazis do?). The protests against CAA began in Jamia Millia Islamia University in Delhi and spread first to Muslim-majority universities and colleges, like Aligarh Muslim University in Aligarh and Nadwa College in Lucknow, but soon universities and youngsters in cities across India were part of it.

These essentially peaceful, non-violent protests were sought to be quashed through police action in BJP-ruled states. This was most heavy-handed in the BJP-ruled Uttar Pradesh (UP), Karnataka, and in Delhi, where the police remain under the direct control of the BJP-run Union Home Ministry. In some cases, especially in Delhi and UP, police officers were captured on video destroying private cars, two-wheelers, shops, and business establishments. In some Muslim-dominated towns of UP, the police entered the homes of the minority community and deliberately smashed up expensive white goods and electronics besides roughing up the occupants.

In Mangalore, in BJP controlled Karnataka, the police opened fire on the protestors, killing two of them. Although by and large the protests were peaceful, a police officer was killed in Delhi, and hundreds of police officers were injured in UP, where 19 civilians were killed and 1,246 arrested during the protests. The UP government invoked the Prevention of Damage to Public Property Act, 1984, imposing community punishments and recovering the loss to public property by seizing and auctioning the property of the protestors. None of this, however, prevented the spread of protests and sit-ins. The most iconic of these was one of Muslim women protesting at Delhi's Shaheen Bagh suburb. It was replicated at hundreds of places in the country where for the first time women from the minority community asserted their citizenship rights.

The Inaction of the Mainstream Parties

The mainstream political parties refused to assume leadership of the anti-CAA protests or provide them organizational backing to sustain their momentum. Except for the Left parties and the Trinamool Congress in West Bengal, the Opposition refused to join the spontaneous public protests, although some cheered from the sidelines. In West Bengal, however, Chief Minister Mamata Banerjee, who enjoys considerable support among Muslim voters (nearly 30 percent of the population), led daily protests against the changes in the citizenship law. Some protests were also organized by Muslim political parties like the Welfare Party and Social Democratic Party in Kerala, and the All India Majlis-e-Ittehadul Musalmeen in Telangana.

In many Opposition-ruled states, the governments facilitated peaceful public protests. The Chief Ministers of West Bengal, Punjab, Kerala, Madhya Pradesh, and Chhattisgarh announced that they would not implement the new citizenship law. In Congress Party-ruled states, the Chief Ministers organized "peace marches" against the CAA/NRC and celebrated December 29, the foundation day of the party, as "Save the Country, Save the Constitution" day.

The largest Opposition party, the Congress, consciously decided not to join the ongoing public protests. Even though one of its most known public faces and party general secretary, Priyanka Gandhi, joined a student sit-in for a couple of hours at Delhi's India Gate, the party decided that its youth wing would organize separate protests instead of joining the ongoing agitations. Apparently, the leaders of the party were wary of religious polarization and in such an eventuality they wanted to protect the party from collateral damage—such as being dubbed pro-Muslim and anti-Hindu. The Aam Adami Party (AAP), once the hope of the Muslim community in Delhi, was also cautious about openly supporting the citizenship law protests. Like the Congress party, it feared being projected by its BJP rivals as yet another party that believes in "minority appeasement."

The political parties that have had a sizeable Muslim constituency also did not actively lead the public protests. The top leadership of the Samajwadi Party, the Bahujan Samaj Party, and the Rashtriya Janata Dal were absent from the protests in North India. It seemed that being pro-Muslim was dangerous political branding even for these parties. The ruling BJP was happy with these developments. It believed that the potential for communal polarization in the anti-citizenship law protests would lead to Hindu consolidation. It was also confident that it had put the Opposition on the back-foot since no one wanted support the Muslim community openly.

Muslim Women Take the Lead

The inaction of the mainstream Opposition parties sent a clear signal to the Muslim community that these parties did not want to engage with it as intensely as they had before the 2014 general election, which had brought the BJP to power. They were even afraid of showing solidarity with them when their citizenship itself was threatened. It was almost as if the triumph of the Hindu right wing populism of the BJP was complete. The Muslim community perhaps came to the realization that although the mainstream parties banked on its votes, they had never given it its fair share either in the party structure or as candidates in legislative politics. Thus, the unsaid compact reached after Independence, that Muslims would be politically represented by mainstream secular parties and work through them, started coming apart. Yet, the Muslim community did not react by either becoming insular or retreating to orthodox Muslim parties. There was no radicalization of the community (except for Jammu and Kashmir for historical reasons and in some parts of Kerala). Nor did it choose to articulate its protest in religious terms.

In the protests against CAA the community seemed to be developing a new vernacular idiom choosing secular, civic symbols over religious identity. Muslim women, some of whom had spoken sporadically about access to places of worship or demanding changes in inequitable personal laws such as instant divorce, were leading public protests and participating in a public discourse about the Constitution and citizenship rights. They held up portraits of the Father of the Nation, Mahatma Gandhi, B. R. Ambedkar (considered the father of the Indian Constitution and Dalit leader), Jyotiba Phule (anti-caste social reformer), Savitribai Phule (wife of Jyotiba and a social reformer, feminist, and educationist), and Maulana Abul Kalam Azad (freedom fighter, votary of Hindu-Muslim unity, and an educationist). They read the Preamble of the Constitution and held up the national Tricolour. Muslim secularism, which took the Constitutional rights of the community for granted, was now being actively asserted by Muslim women. Many said that they were doing this to secure the future of their children.

The platform they offered to speakers at protest sites like Shaheen Bagh was tightly controlled, with those straying into religious arguments being told to stick to Constitutional and secular issues. The leaders of mainstream political parties who came to address the women's protests had to abide by the terms set by the protestors.

This also became a unique moment of Muslim women coming together for a political purpose. Housewives mingled after completing domestic duties with women who joined in after office hours and women students at the sit-in. Muslim women who had not come out to protest, even when men from their community were lynched by cow vigilantes, were out day after day in large numbers. Shaheen Bagh caught the imagination and spawned several protests led by Muslim women all over the country where they were joined by men and women of other religious and political persuasion. Curiously, there were no clearly identifiable leaders. The protest sites were managed by committees of volunteers. They spoke about the limited perspectives of all traditional political parties and their inability to address emerging issues such as CAA, Constitutionalism, and even gender disparities. Shaheen Bagh became a model of broad political accommodation with even the LGBT community being given the space to speak. If Muslim women were no longer apologetic about their identity, they also wanted others to be equally unapologetic. Identitarian affiliations could not be allowed to question their patriotism and nationalism. Protest venues also became sites for socializing and community interactions long into the night. Many at the protests felt that it was an advantage to not have any clear leadership—it allowed everyone to express an opinion, all views were equally important, and everyone was seen as a potential leader. In the absence of identifiable leaders, the inimical state machinery could not malign the protests by linking them to one leader or the other. Many activists thought that it was important to recognize that a politicization process was taking place in the Muslim community; that the citizenship issue was only an immediate cause of the protests but the movement would also help focus attention on issues of gender inequality, patriarchy, caste inequality among Muslims, and the poverty of the community. Only a leadership that could appreciate these issues, they felt, could take the potential of the protests forward.

Even if a new leadership does not crystallize out of these protests in the Muslim community, they have created a moment of political self-awareness, which will have a lasting impact on the consciousness of the Muslim community, especially its women. It is unlikely that they will once again put their faith in the traditional political leadership, orthodox or secular. Whatever new leadership emerges is likely to be rooted in constitutionalism. The solidarity that the protests built up with other communities, such as Sikhs and Dalits for example, could sow the seeds of its second phase. However, the process of sensitizing the middle class and lower middle class Hindus to the fact that the new citizenship initiatives affect not only Muslims but also poor Hindus, tribals, and Dalits, was cut short by the coronavirus pandemic. It is difficult to say what form the protests will take after the pandemic subsides, but there is little doubt that they will revive in some form should the BJP continue with divisive policy on citizenship.

Sabyasachi Basu Ray Chaudhury

Dispossession, Un-freedom, Precarity:
Negotiating Citizenship Laws
in Postcolonial South Asia

. . . no one puts their children in a boat
unless the water is safer than the land
no one burns their palms
under trains
beneath carriages
no one spends days and nights in the
stomach of a truck
feeding on newspaper unless the
miles travelled
means something more than journey
no one crawls under fences
no one wants to be beaten
pitied . . .
—*Home*, Warsan Shire

The partition of the Indian subcontinent forced millions of people to flee to
the other side of the borders, freshly demarcated by the British colonial rulers just on the eve of their hurried departure from South Asia. Almost a
decade-long migration of people could not, however, settle the boundaries
and lives of the people once and for all. The postcolonial rulers retained
many of the draconian laws of the late colonial period, like the Foreigners'
Act, 1946, and laced them with new laws and regulations, thus leading to
greater dispossession of people of homes, generating widespread situations

The South Atlantic Quarterly 120:1, January 2021
DOI 10.1215/00382876-8795866 © 2021 Duke University Press

of un-freedom, and creating countless refugees, stateless persons, and even migrants, mostly forced to survive in sites of precarious life, without any right to have rights. When the world is in the midst of a pandemic, facing mixed and massive flows of population, the people living in precarity are almost in similar situations. Whereas citizenship of a nation was supposed to ensure the rights and privileges of people inhabiting a territorial space, demarcated by political boundaries, laws in postcolonial South Asia have led to exclusion of many people from citizenship on account of their religion, ethnicity, or gender, thereby practically turning into mechanisms of newer politics of dispossession.

In this essay, we intend to explore, in brief, the relationship between precarity and citizenship laws in South Asia as the rise of precarity is fueled through dispossession of women, and ethnic and religious minorities of their rights and entitlements, thus pushing them toward a world of un-freedom. The migrant laborer also suffers from a double bind of working in inhospitable conditions, and of being exposed to precarity on account of their lack of citizen rights. We shall underscore the close brush of these dispossessed people with un-freedom by examining the citizenship laws of India, Myanmar, Pakistan, Bhutan, and Sri Lanka, as they have unfolded in postcolonial times.

Decolonization of India and After

When the British colonizers left the subcontinent in 1947 in a hurried manner, exactly 190 years after the Battle of Plassey, and after eighty-nine years of direct rule in the name of British monarchy, two separate states—India and Pakistan—emerged out of partition, ostensibly along religious lines, the former being a Hindu-majority nation, and the latter a Muslim-majority one. As the people living in the region started apprehending a catastrophic partition, their attempt to escape this led to one of the largest mass migrations in modern times. At least fifteen million people were compelled to migrate and perhaps about two million innocent men, women, and children were butchered to death. Ayesha Jalal (2013: 4) rightly mentions that, this "partition continues to influence how the peoples and states of postcolonial South Asia envisage their past, present and future" (see also Singh 1988; Khan 2008). Over and above, the Bangladesh War of 1971 triggered a fresh wave of migration to India, this time of people of both Hindu and Muslim origins.

The colonizers left, but their canons of controlling people remained almost intact. Fresh rules, curtailing freedom, were very often thrust upon people even in the newly-born nations of South Asia experimenting with democracy. Therefore, the colonial laws governing the inflow of foreign nationals into India, like the Passport (Entry into India) Act, 1920, Registration of Foreigners Act, 1939, and Foreigners' Act, 1946, have been supplemented by other draconian laws in postcolonial times. Even though India has received the largest number of refugees in South Asia since her independence, New Delhi has stopped short of institutionalizing the practices already adopted by it. As refugees are treated as "foreigners" under the existing Indian laws, claims for refuge are mostly considered on a case-to-case basis. For the lack of any codified refugee law or policy in India, refugees are usually treated on the basis of their ethnic and/or religious identity, and country of origin, and they very often turn into "strategic pawns" in the chessboard of Indian diplomacy.

Scheming the Refugees

India is not a signatory to the 1951 UN Convention on Refugees and 1967 Protocol. But, it is a signatory to a number of international human rights laws and international humanitarian laws, like UDHR (Universal Declaration of Human Rights), 1948, ICCPR (International Covenant on Civil and Political Rights), 1966, and ICESCR (International Covenant on Economic, Social and Cultural Rights), 1966. Therefore, although according to these international legal instruments, India is supposed to ensure the basic rights of all persons, citizens and non-citizens alike, the lack of consistency in treatment of refugees in India persists. For example, when the government of neighboring Myanmar started suppressing the pro-democracy movement there in the late 1980s, and hundreds of citizens of Myanmar sought refuge in India, New Delhi assured those displaced people that no genuine refugee would be turned back, as at that point in time, India, at least covertly, was critical of steps taken by Yangon, and was sympathetic to the pro-democratic movement in Myanmar.

However, when subsequently thousands of innocent men, women, and children belonging to the Rohingya community were stripped of their citizenship and practically expelled from Myanmar, in view of the wanton destruction of their home and property following the adoption of the Citizenship Act of 1982, and some of these stateless Rohingyas sneaked into Indian territory for shelter, they were treated as "illegal immigrants," "foreigners," and a potential risk to India's national security.[1] After all, in the early 1990s, India made a major shift in her policy vis-à-vis Myanmar's ruthless military junta in the context of growing Chinese influence in the region. In other words, the Rohingya refugees in India became a victim of India's Look East Policy and Act East Policy, and of New Delhi's developing strategic interests in South and Southeast Asia.

In the context of this practice of selective recognition and acceptance of refugees by India, the humanitarian concern of the new Citizenship Amendment Act (CAA) of India, 2019, is primarily applicable to the Hindus, Buddhists, Parsis, Sikhs, Jains, and even Christians from Pakistan, Afghanistan, and Bangladesh, thereby excluding thousands of Tamil refugees, mostly Hindus, fleeing the sites of civil war and genocide in Sri Lanka, hundreds of Rohingya, mostly Muslims, fleeing ethnic cleansing in Myanmar, and Muslims of Shia and Ahmadi stock, facing persecution in Pakistan. In other words, the inflow of Hindus to India is frequently considered as a "natural homecoming," while that of Muslims, even if they are refugees from Myanmar, or any other neighboring country, may be treated simply as "infiltration." This tactical omission has generated a lot of debates and given rise to agitations in different parts of the country, and the Supreme Court of India is yet to decide the constitutionality of the CAA.

In fact, so far, a number of judicial decisions in India have ruled that, "well-founded fear," the catchword of identifying a "refugee" according to the 1951 UN Convention, needs to be backed by the evidence of "real and substantial danger of persecution" for someone to gain refugee status in India. The Indian Evidence Act, 1872, after all, puts onus on the claimant to prove the threat of persecution. Under the circumstances, New Delhi did not have any hesitation to deport a number of Rohingyas to Myanmar in 2019, even as thousands of Rohingyas fled a fresh spell of violence in their country of origin in that year.

Meanwhile, the updating of the National Register of Citizens (NRC) in Assam in August 2019, to determine citizenship of "suspicious" residents of this state in India's Northeast, has led to the exclusion of 1.9 million people, both Hindus and Muslims, already residing there, probably for decades. This heinous attempt to consolidate the vote of religious majority in the country's troubled frontiers, in the long run, is likely to complicate interethnic and interreligious relations in a region that was governed by the British colonial rulers more often than not through the "Inner Line Permit" (ILP) to regulate the movement of "outsiders" in the "protected areas," mainly inhabited by different indigenous communities that could not be brought under adequate control of the colonizers.

Perils in Pakistan

Similarly, the Pakistan Citizenship Act, 1951 is considered to be highly discriminatory toward women. This issue has come to the fore in recent times in view of the problems relating to Afghan refugees in Pakistan. Many Pakistani women wedded to Afghan men expressed their concern over the growing pressure from law-enforcing agencies, including police, on Afghan refugees to return to their own country (Shah 2016). But, some of these families are in a tight spot as Pakistan's citizenship law only permits a foreigner spouse of a Pakistani man to get Pakistani citizenship. The same facility is not meant for the foreigner spouse of a Pakistani woman.

In this situation, the Federal Sharia Court took a *suo motu* notice of this issue of discrimination in the citizenship law. A bench comprising Chief Justice Haziqul Khairi, Justice Fida Mohammad Khan, and Justice Salahuddin Mirza, after hearing the federal government, the National Commission on the Status of Women, provincial governments, and Aurat Foundation, said in its verdict on December 12, 2007, that: "We are of the view that Section 10 of the Citizenship Act is discriminatory, negates gender equality and is in violation of Articles 2-A and 25 of the Constitution of the Islamic Republic of Pakistan and most importantly is repugnant to the holy Quran and Sunnah" (Shah 2016). However, it added that: "In the end, it may be clarified that while Section 10 of the Citizenship Act 1951 expressly contemplates denial of citizenship to a foreign husband of a Pakistani woman as neither she or he is entitled to apply under it, nevertheless under law whether municipal or international, grant of nationality would remain within the domain of discretion of the government of the country which may refuse it for reasons of national security or public interest, etc. to which she or he cannot claim it as a matter of right" (Shah 2016). Finally, the bench said: "We . . . in the exercise of our powers under clause (3)(a) of Article 203–D of the Constitution require the President of Pakistan to take suitable steps for amendment of Section 10(2) and other provisions of the Citizenship Act, 1951, within six months hereof ensuring appropriate procedure for grant of Pakistani nationality to a foreign husband married to a Pakistani woman" (Shah 2016).

However, subsequently the interior ministry of Pakistan stated that foreign women marrying Pakistani men could not be equated with foreign men married to Pakistani women in the society. It said: "It would add legal stay of a large number of illegal immigrants residing in Pakistan and also increase their influx" (Shah 2016). It further added that a "majority of persons would misuse the provision especially illegal immigrants like Afghan refugees,

Bengali, Behari" and other communities of South Asia "who do not intend to return to their country" (Shah 2016). It finally cautioned that "apart from social/economic implications, the provision can also be used by any foreign country to plant their agents in Pakistan." It even opined that "the proposed amendment may be assessed on the touchstone of Indo-Pak relations; it will provide legal ingress to Indian male citizens into Pakistan" (Shah 2016).

Banished from Bhutan

The people of Nepali origin started migrating to Bhutan in the nineteenth century, and gradually began becoming citizens of their adopted country. However, by the late 1970s, the dominant Drukpa establishment identified the growing number and influence of ethnic Nepalis as a threat to Bhutan's "cultural identity" and to the privileged position enjoyed by Drukpas in the Bhutanese society. Therefore, the Lhotsampas, or people of Nepali origin, were gradually identified as "illegal immigrants" who appeared to threaten Bhutan's "survival as a distinct political and cultural entity" (Ministry of Home Affairs 1993: 41). These threats were invoked by the government to have two new Citizenship Acts in Bhutan, in 1977 and 1985 respectively, which altered the conditions drastically for acquiring citizenship of the small Himalayan nation.

For instance, the Citizenship Act, 1977 increased the residency requirement for citizenship from five to fifteen years for government employees, and from ten to twenty years for the other foreigners. There was also a requirement of having "some knowledge" of Dzongkha, the language of the Drukpas, and of Bhutanese history. Over and above, it was stated categorically that citizenship would not be accorded to anyone involved in the activities against the people, the country, and the King. Bhutan's first census subsequently facilitated to identify "citizens" and issue identity cards only to them. Later on, another census was conducted in 1988 after the promulgation of the Citizenship Act, 1985. Interestingly, the second census was conducted primarily in southern Bhutan, an area inhabited largely by the Lhotsampas. The government even refused to recognize the citizenship identity cards issued by it after the previous census, and therefore, such people belonging to the Lhotsampa community were identified as non-nationals or "illegal immigrants." The ultimate step in the Bhuatanization policy perhaps was the introduction of *Driglam Namzha* (Thronson 1993: 20), the traditional Drukpa code of values, etiquette, and dress, in January 1989, through a decree issued by the King of Bhutan. In the very next month, the

Nepali language was taken out of the school curriculum in southern Bhutan. Against this background, by 1990, the Government of Bhutan compelled "non-national" Lhotsampas to leave the country, after forcing them to sign "voluntary migration" forms (Amnesty International 1994), and most of these people of Nepali origin, more than 120,000, had to take refuge in Jhapa and Morang districts in eastern Nepal after they trekked to those areas through the adjacent state of West Bengal in India.[2]

Inclusion as Exclusion

Sri Lanka became infamous for disastrous consequences of her exclusionary citizenship legislation within months of Ceylon (the former name of Sri Lanka) being decolonized in February 1948. Although the Soulbury Constitution of Ceylon, drafted primarily by the British rulers, perhaps had the intention to turn their "model colony" into a "model democracy," the Citizenship Act, 1948 was the beginning of a string of exclusionary tactics of the majority Sinhala leadership to consolidate their political base immediately after the decolonization of the island nation. This act was primarily aimed at the Tamils of Indian origin in Sri Lanka, the major workforce in the up-country tea plantation. The Citizenship Act clearly delineated ethnic divide in the island nation, and altered the political system in favor of the Sinhala majority. Under this act, citizenship could only be secured through registration or patrilineal descent. To register oneself as a citizen, the unmarried applicants were to show at least ten years of uninterrupted stay in Sri Lanka from the date of the application. The married ones required at least seven years of continuous stay. To produce evidence of such stay was a tall order for the poor and mostly illiterate plantation Tamils.

The problem of plantation Tamils, numbering about a million at the time of the country's decolonization, was attempted to be solved through the Sirimavo Bandarnaike-Lal Bahadur Shastri Pact, 1964. This pact between Sri Lankan and Indian Prime Ministers came in the aftermath of the Sino-Indian Border Conflict of 1962, when bruised New Delhi became more eager to mend fences with Colombo. Therefore, it was readily agreed upon by India and Sri Lanka that New Delhi would accept 525,000 Tamils of Indian origin and their natural increase, while Colombo would accept 300,000 and their natural increase. The status of the remaining 150,000 was to be decided later. However, most of these plantation Tamils were reluctant to leave Sri Lanka after staying there for three generations or more since the British colonial period. About 400,000 plantation Tamils applied for the citizenship of India and 630,000 applied for that of Sri Lanka. In reality, only 162,000 were granted the Sri Lankan citizenship and a little more than 350,000 got the Indian citizenship. The Tamils remaining in Sri Lanka overnight became stateless (Subramanian 2020).

Finally, we shall examine how the specter of COVID-19 pandemic could in no time upset the life of the "invisible" migrant labor within and from South Asia, suddenly catapulting their everyday precarity to the center stage of discussion.

Pandemic and After

The capitalist globalization of recent times has facilitated the mobility of capital, and even technology, but not the mobility of labor. Therefore, the condition of migrant labor within the country and abroad today are in no way better than that of the refugees, asylum-seekers, or stateless persons. COVID-19 pandemic has only foregrounded the precarity, un-freedom, and dispossession of migrant labor across the globe.

Every year, millions of people from South Asia go mainly to the countries of the Persian Gulf region in search of a livelihood. More than 8.5 million migrant laborers from India were in the Gulf States in 2018 according to the data of the Ministry of External Affairs, Government of India. Kerala, a southern state in India, is one of the largest migrant-sending states of the country and one of the biggest beneficiaries of the "money order economy" (meant to describe remittances from migrant labor) in South Asia. Migrants send money to their families to boost their living standards. These remittances to their home economy undoubtedly enrich the exchequer of countries like India and Bangladesh. For example, according to World Bank statistics, in 2019, India received US $82 billion as remittance (Budhathoki 2020).

However, the outbreak and global spread of the COVID-19 pandemic have put tremendous pressure on these migrants. For example, Qatar's forceful and illegal deportation of Nepali workers in March 2020 under the pretext of testing them for the coronavirus has raised new fears about the fate of South Asian migrant workers in the Middle East (Budhathoki 2020). These suddenly expelled workers did not even have an opportunity to collect their belongings, pay, or other benefits, and also did not have the wherewithal to challenge their expulsion. News reports indicate that the migrant workers were detained in Industrial Area, Barwa City, and Labour City, all of which are parts of the capital Doha, and were informed by the police that this was for a medical checkup (Budhathoki 2020). Afterwards, the Qatar police collected their biometric information and crammed them into makeshift detention centers, before putting them in overpopulated accommodation without even providing adequate food and water.

Most of these migrant workers from India, Bangladesh, and Nepal were engaged in building infrastructural facilities for the 2022 FIFA World Cup, and were, in any way, facing poor living conditions and unsafe workplaces in the Doha Industrial Area, quite infamous for its slums and overcrowded labor camps. Most of the Gulf countries, flush with petrodollars, build their fortresses with the blood and sweat of migrant labor surviving in the most inhuman state. The outbreak of pandemic has made migrant laborers extremely vulnerable as they are viewed as the source of infection (Budhathoki 2020).

Even within India, the "total lockdown" imposed by the federal government through announcement of the Prime Minister within a very short notice, to prevent spread of the deadly virus, has come as a double blow for hundreds and thousands of migrant laborers of Uttar Pradesh, Bihar, West Bengal, and Odisha, living in slums and small rented accommodation in Delhi, Maharashtra, or Gujarat, where thousands of families are huddled together.[3] On the one hand, lockdown has jeopardized their livelihood opportunities completely, at least for the time being, rendering them almost penniless, and on the other, the advice of ensuring "social distancing" and washing their hands with soap and/or alcohol-based sanitizer frequently is none other than a cruel joke, where dozens of people have to share the same source of water, or where ten, fifteen, or twenty persons have to live in a small room.

These migrant workers, who toiled in small factories, shops, or as courier or delivery persons, mostly have not received their salary ever since the lockdown was clamped within a few hours. They wanted, therefore, to return to their home states. As no transport was available, and all the state borders were sealed, people died on their unusually long trek of 250–500 kilometers back home due to lack of food and safe drinking water, and out of exhaustion in rising summer temperatures of north, central, or west India (Chaudhury 2020). While the Government of India arranged international flights to bring back the nation's elite, white collar workers, and tourists stuck in other corners of the world, the poor and less-skilled migrant laborers within and outside the country were mostly left high and dry. This brings to the fore a new class division existent among the migrants. That is why Slavoj Žižek (2020: 26) raises the pertinent question: "what about those whose work has to take place outside, in factories and fields, in stores, hospitals and public transport?" And, he also provides the response in no time: "Many things have to take place in the unsafe outside so that others can survive in their private quarantine" (Žižek 2020: 26).

On another plane, the Malaysian government has recently risked lives by pushing back overloaded boats of Rohingya refugees in these times of pandemic as it blocked lifesaving rescues of seaborne asylum-seekers. It has pushed back to sea at least two boats filled with Rohingya refugees. On April 16, 2020, the Malaysian navy intercepted a boat with around two hundred Rohingya refugees off the coast of Malaysia and prevented the boat from entering Malaysian waters. The previous day, Bangladesh Coast Guard officials intercepted another boatload of refugees that had been turned away from Malaysian waters almost two months earlier, cocking a snook at the principle of *non-refoulement* (Human Rights Watch 2020) emphasized in the international refugee laws.

In short, the pandemic has further exposed the precarity of refugees, asylum-seekers, and stateless persons, along with migrants, including the undocumented ones like never before. Their choice is limited between starvation and disease, and these heart-wrenching stories of human plight should worry us all in the world even after the pandemic, which has practically been identified as "disease of migration" (Srinivasan 2020). We may emerge in a world with a "new normal," where right to work, social relationships, and even friendships and ideas of hospitality are likely to change beyond recognition after a war with an invisible enemy. Be that as it may, the COVID-19 pandemic is likely to leave us with the legacy of more suspicion, exclusion, mistrust, stigma, and precarity in view of the dominant norms of citizenship in South Asia, having the potential for generating more stateless people in the years to come.

Notes

1   For details see Chaudhury and Samaddar 2018.
2   For details see Hutt 2005; Refugee Watch 1998.
3   For details see Samaddar 2020.

References

Amnesty International. 1994. *Bhutan: Forced Exile.* AI Index ASA, April 14.Budhathoki, Arun. 2020. "Middle East Autocrats Target South Asian Workers." *Foreign Policy*, April 23. foreignpolicy.com/2020/04/23/middle-east-autocrats-south-asian-workers -nepal-qatar-coronavirus/.

Chaudhury, Sabyasachi Basu Ray. 2020. "Notun Osukh, Purono Baishomyo (New Disease, Old Discrimination)" (in Bengali). *Anandabazar Patrika*, April 3. anandabazar.com /editorial/coronavirus-lockdown-workers-walks-down-300km-just-to-return-home -1.1130827#.XoaXrJ137cg.whatsapp.

Chaudhury, Sabyasachi Basu Ray, and Ranabir Samaddar, eds. 2018. *Rohingya in South Asia: The People without a State*. London: Routledge.

Human Rights Watch. 2020. "Malaysia: Allow Rohingya Refugees Ashore: Covid-19 No Basis for Pushing Back Boats." *Human Rights Watch*, April 18. hrw.org/news/2020 /04/18/malaysia-allow-rohingya-refugees-ashore.

Hutt, Michael. 2005. *Unbecoming Citizens: Culture, Nationhood, and the Flight of Refugees from Bhutan*. New Delhi: Oxford.

Jalal, Ayesha. 2013. *The Pity of Partition: Manto's Life, Times and Work across the India-Pakistan Divide*. Princeton: Princeton University Press.

Khan, Yasmin. 2008. *The Great Partition: The Making of India and Pakistan*. New Haven: Yale University Press.

Ministry of Home Affairs, Government of Bhutan. 1993. *The Southern Problem: Threat to a Nation's Survival*. Thimphu: Government of Bhutan.

Katel, Narayan. 1998. "Exiled from the Kingdom." *Refugee Watch*, no. 3 (July).

Samaddar, Ranabir, ed. 2020. *Borders of an Epidemic: COVID-19 and Migrant Workers*. Kolkata: Calcutta Research Group.

Shah, Waseem Ahmed. 2016. "View from the Courtroom: Pakistan's Citizenship Law in the Limelight." *Dawn*, September 19. dawn.com/news/1284651.

Singh, Khushwant. 1988. *Train to Pakistan*. New Delhi: Orient Longman.

Srinivasan, Aditya. 2020. "Migrant Flows Will Change, They Will Take Time to Come Back to the City, Says Scientist S Irudaya Rajan." *Times of India*, May 3. https://timesofindia .indiatimes.com/india/migrant-flows-will-change-they-will-take-time-to-come-back -to-the-city-says-scientist-s-irudaya-rajan/articleshow/75510677.cms.

Subramanian, Nirupama. 2020. "Exclusion and Ethnic Strife: Story of Sri Lanka's Citizenship Law." *Indian Express*, January 17. indianexpress.com/article/explained/exclusion -and-ethnic-strife-story-of-sri-lankas-citizenship-law-6218590/.

Thronson, David. 1993. *Cultural Cleansing: A Distinct National Identity and the Refugee from Southern Bhutan*. Kathmandu: INHURED International.

Žižek, Slavoj. 2020. *Pandemic! COVID-19 Shakes the World*. New York: Or Books.

Sanjay Barbora

Counting Citizens in Assam: Contests and Claims

The superfast Rajdhani Express train stops for less than two minutes in western Assam's Kokrajhar station. The brief halt is usually a hurried one as day travelers jostle each other to catch the train to the capital city Guwahati. However, on March 15, 2020, with the possibility of a lockdown and government-sponsored information to remain home, the station was almost empty as I got on with another person. Both of us wore masks and found ourselves in a compartment that was nearly empty, a rare occurrence on a train that makes a long 2,300–kilometer journey from New Delhi to Dibrugarh in Assam. "People are afraid of this corona, so they are not traveling," he said as he went on to explain that he had been invited to a local college in Kokrajhar to receive an award for his many innovations in machinery. A middle-aged person from the eastern part of Assam, he had spent all his adult life tweaking and twirling machines to ensure that they could be built at a low cost and for people who were engaged in various kinds of agriculture. By then, the idea of a self-declared, self-imposed quarantine for travelers from abroad was being considered and debated. "I would feel uneasy declaring anything to the administration, you know," he said as we were offered complementary tea on the train. "They will lock us up, just like they want to with the illegal immigrants in the detention centers . . . so, I would rather stay home quietly and not disclose any symptoms," he added pursing his lips, as if to announce that we ought to speak about other matters.

For me, this fortuitous encounter reveals the outcome of several decades of militarization and political mobilization in Assam that implicate both the state and those seeking to challenge it. It speaks about the inversions that happen when governments appropriate selective portions of dis-

The South Atlantic Quarterly 120:1, January 2021
DOI 10.1215/00382876-8795878 © 2021 Duke University Press

senting narratives in order to perpetuate their control over resources and territory. The recent interest and focus on citizenship debates in Assam (and India, more generally) were upended by the Novel Coronavirus pandemic. However, in the course of this essay, I draw on three important factors that have contributed to the current predicament, where the Indian state and its citizens in Assam display different ideas about what constitutes an association of equals. They are: (a) the tensions between autonomy and social justice in Assam; (b) the history of counter-insurgency and governance in the region; and (c) the emergence of software and technology as arbiters of an old twentieth-century question that continues to animate politics into the first quarter of the current one.

Together, they constitute an apparatus of the Foucauldian kind, where the network created between the three serves to illuminate juridical, technological, and military power relations within society (Agamban 2009). In looking at the manner in which the apparatus came to be established, one is able to make sense of a paradox where the state that claimed to have contained insurgency continues to treat sections of its citizens as potentially dangerous. Indeed, since the detention of rebel leaders and commencement of peace talks between them and the government, one can argue that political violence between the state and insurgents has declined in Assam since 2010–11. However, this has not nurtured a milieu where communities have been able to air grievances and resolve old disputes. Instead, the violence has been directed against neighbors, a process that has been abetted by sustained political campaigns against immigrants. Therefore, as Anupama Roy (2002: 29) pointed out, when the government of India amended the country's citizenship laws in 2003 it "emphasised the wall of separation between citizens and non-citizens by inserting in the section on citizenship by birth, the distinction between those who were born to Indian parents and were Indian citizens through descent and blood ties, and those who could not make such claims to citizenship by birth." This fostered the creation of an adversarial environment within Assam, where an enduring fault line about immigration was heightened at the cost of other, equally pressing political concerns.

Autonomy and Social Justice

Citizenship debates in Assam predate Indian independence and bring together two concerns that mark most histories of colonization for those at the receiving end: social justice and autonomy. The colonial period is a key to many of the enduring conflicts in Assam today. Adversarial positions on the

National Register of Citizenship (NRC) that ended in August 2019 fall into a process that has been researched and documented well over the past few decades. The presence of the colonial state in Assam was limited to parts of the populated valleys, where the government allowed people from erstwhile East Bengal to settle on agricultural land for annual and decennial leases. The landscape, economy, and society changed dramatically, as cash crops like jute and tea, as well as minerals like oil and coal, were grown or extracted in abundance from the area in the nineteenth and early twentieth century. This transformation also entailed a radical change in the demography of the region, as peasants and indentured workers from different parts of the British-controlled Indian subcontinent were brought to Assam. Tea plantations, in the central and eastern part of the Brahmaputra Valley and in parts of the Barak Valley, were given long-term leases.

In the upland areas, however, the government followed a 'light-touch' policy and allowed indigenous communities to retain their traditional chiefs and heads, while making way for indirect rule by the colonial state. This policy continued after independence and was reaffirmed by the Bordoloi Commission in 1949, when they proposed that the hills be governed under the Sixth Schedule of the Indian Constitution. Under the provisions of the Sixth Schedule, use and transfer of land between individuals was left to the discretion of the autonomous councils that allowed indigenous communities (defined as Scheduled Tribes under the Indian Constitution) to govern certain areas where they were a numerical majority. The councils functioned as territorial enclaves within the larger state, and in matters related to transfer of land and property, reflected the light-touch administration during the colonial period. While some territories and communities accepted this autonomy arrangement, others like the Naga and Mizo were less convinced. In both areas—Naga Hills (comprising the current state of Nagaland and parts of Arunachal Pradesh, Assam, and Manipur) and Lushai Hills—demands for independent, self-governing territories brought together small, kin-based communities who were able to organize successful armed resistance to the postcolonial state and to settler communities. The first territorial councils were elected in the autonomous districts around the undivided province of Assam in the 1950s and continue into contemporary times. Since then, the state of Assam has been reorganized, and currently there are three territorial autonomous district councils (Bodoland, Dima Hasao, and Karbi Anglong) and six non-territorial councils (Deori, Mishing, Rabha Hasong, Sonowal Kachari, Thengal Kachari, and Tiwa) in the state today.

Given the region's complex history of immigration, especially the one that began with the establishment of colonial rule and tea plantations in the late nineteenth and early twentieth century, this constitutional blind spot has had very tragic consequences for certain groups of people. This is especially true of those whose identities are tied to the kind of labor they do in the plantations and paddy fields. Adivasi activists and advocacy groups among Muslims of East Bengal heritage in Assam have had to constantly remind political organizations and civil society groups about their contributions to the region's culture and social history (Kikon 2017; Azad 2018). Today, the Adivasi communities in Assam are joined by five other communities in their demand to be included among the Scheduled Tribes in the state. This is a crucial aspect of their political struggle, as it alludes to some basic, watered down acceptance of reparations for historical injustices of the tea industry's indentured system of labor recruitment. These tensions between demands for autonomy (exemplified in demands for homelands) and social justice (as exemplified by demands for inclusion as rights-bearing citizens) converged in the government's violent response to popular demands around these issues.

Counterinsurgency and Governance

The army has been a significant presence since its inception into civic life and politics in Assam in 1990. For over two decades, military operations have resulted in the creation of a climate of impunity and human rights abuse by the state. Counterinsurgency was a pervasive aspect of political life, to the extent that those involved in it became active stakeholders in public life, as well its shadowy counterpart that involved grey areas of commerce, politics, and other unrestricted matters (Baruah 2019). Indeed, one of the main concerns of a colonial counterinsurgency regime has been to restore the legitimacy of the government, while continuing to fight a military war against sections of the population. Robert Thompson's (1966) detailed treatise on conducting counterinsurgency campaigns includes the need for isolating guerrillas from people, assuming that much of the population would remain neutral and uninvolved in the initial stages of the conflict. He emphasizes that the government would have to draw on three aspects in the long run: (a) nationalism and national policies; (b) religion and customs; and (c) material well-being and progress.

Ever since 1990, these elements had become enmeshed in the manner in which civil administration addressed issues of governance. The Unified

Command structure was introduced in 1990, and even though the Chief Minister and Chief Secretary were placed at the apex of the decision-making process, the actual groundwork was conducted by the army. This extended to the highest administrative structures in the 1990s and early 2000s, when most of the governors of the state were retired military generals, who were able to engage more closely with their former colleagues in order to encourage greater connection with the Indian mainstream (Baruah 2001).

Nationalist slogans and policies that were introduced in the early 2000s were instrumental in bringing about a significant change among Assam's middle classes, especially regarding their misgivings about human rights violations committed by the state (Deka 2019). Additionally, ever since the late 1990s, civil administration in many districts had grown accustomed to religious congregations of the three main faith-based communities in the state. As a PhD researcher in the autonomous district of Karbi Anglong in 2003–2004, I would cross a Christian evangelical healing center and an Islamic Ijtema on my way to an interview at the tribal development office (run by radical Hindu nationalists), all within the span of an hour. At each stop, preachers would try to persuade their flock to either strengthen the community against outside influence, aspire for a better material life, or both. Significantly, it was the Hindu radical nationalists who managed to do developmental work among indigenous (tribal) communities, while mobilizing Bengali Hindu settlers to the cause of a larger Hindu nation (Rajkhowa, Phukan, and Boruah 2018).

This political sifting of communities was perpetuated through the concept of garrisons, where two worlds coexisted uneasily. One was safe, secure and sanitized to include all that was representative of India as a country, where army personnel and their families lived in harmony. The world outside the garrison, on the other hand, was that of the civilian, who were racially different and constantly at war among themselves, as well as with the nation (Barbora 2016). Racial and ethnic superiority has long been a standard deployment in counterinsurgency techniques across the world, since it allowed the normalization of social differences and structural violence (Drohan 2017). The violence outside the garrisons in Assam was aimed at the erosion of solidarity among communities. In fact, it enabled every group to feel as though they had suffered alone and more than their neighbors. An unfortunate outcome of this form of militarization was the erosion of shared communitarian spaces and relationships that had earlier allowed for nonviolent coexistence in the state (Barbora and Sharma 2016).

One of the military outcomes of counterinsurgency was the stalemate created by army and police operations against insurgents. The losses during the conflict were disproportionately high among the insurgents and their sympathizers. Since 2003, the government began to follow two distinct strategies to contain insurgent movements. With some, like the Bodo Liberation Tiger Force (BLTF), it entered into suspension of operations and created political conditions for allowing the former armed groups to enter into mainstream politics. Hence, the BLTF–government ceasefire facilitated the creation of the Bodoland Territorial Council (BTC) that was spread over four districts, officially called the Bodoland Territorial Region (BTR). With others, like the United Liberation Front of Assam (ULFA), the government continued with a combination of military operations, while also ensuring that leaders based outside the country were pushed back into the government's custody. These two processes have carried on since 2003 (with the ceasefire with the BLTF and creation of the BTR) and came to an informal end in 2009–2010 (with the detention of ULFA leaders and beginning of peace talks with them). However, this did not lead to an end to adversarial politics in Assam. Instead, with the filing of a public interest litigation for conducting of the National Register of Citizenship in Assam in 2012, various organizations and political parties in Assam were induced to revisit an old question about citizenship that had been symptomatic of the late twentieth-century politics. With the NRC, the old counterinsurgency idea of community surveillance by the authorities received a new impetus and focus, even as relationships between groups had become much more contested and fractious.

Software, Technology, and the Counting of Citizens

The 2015 edition of the NRC required individuals to show their legacy data that included having a family member's name in the 1951 NRC and/or having the individual (or a direct family member's name) included in the electoral rolls as of March 24, 1971, a day after the Bangladesh Liberation War was formally announced. In case a person was unable to find her or his name in the legacy data, the administration allowed for twelve other documents that could be shown as evidence, provided they were granted before March 24, 1971. These were: (i) land tenancy records; (ii) Citizenship Certificate; (iii) Permanent Residential Certificate; (iv) Refugee Registration Certificate; (v) Passport; (vi) LIC Policy; (vii) Government-issued License/Certificate; (viii) Government Service/Employment Certificate; (ix) Bank/Post Office Accounts; (x)

Birth Certificate; (xi) Board/University Educational Certificate; (xii) Court Records/Processes. These documents have an aura of middle-class respectability to them. They attest to a person having ownership of property, access to education, jobs, and documents that allow her or him to travel at will. However, many itinerant working people—who constitute Assam's unorganized labor sector—were unable to produce these documents. Every person had to take these documents to the nearest NRC Seva Kendra, a government building designated for the purpose gathering and loading the documents on to a database. For many trying to register their names in the database, this was their first encounter with computers and information technology. Hence, they were also daunted by the finality of the exclusion when it happened, since there seemed to be no human error, or authority that could be chastised for failures. As a bureaucratic network of officials and staff, the NRC process was grafted onto the local administrative structure of governance as a time-bound project that had external support for a specific period of time. This allowed the state coordinator of the NRC to deploy various officers (and offices) to ensure a smoother functioning in places where human interface with potential technological shortcomings seemed inevitable.

In most cases, the diligence of individuals and groups meant that most people were able to provide documentation. Years of activism had built into the system some checks and balances that could ensure a process of redress, as well as a human interface that people could appeal to. Therefore, when one's name was excluded from the NRC, a person could appeal to a Foreigners Tribunal (FT), a body that was originally constituted by the government in 1964 and later amended in 2019, specifically to deal with the cases that had come up after the exclusions of 2018. With support from the central government, the government of Assam recruited one thousand members (as those judging the cases under the FT are called) into the FTs. Any advocate between the ages of forty-five to sixty was eligible to apply and would be considered for a contract of two years, where they would be employed by the government of Assam. An interesting, but leading criterion for their recruitment was stated early on, as the government notification stated that the person (applying) should "have a fair knowledge of the official language of Assam and its (Assam) historical background giving rise to foreigner's issue." [1] Such conditionality has severe consequences for people who do not speak standardized Assamese, who are *char*—seasonal river islands—dwellers, and who are itinerant, as they are already seen to be interlopers in the region. They were also most likely to be summoned to the FT members to answer questions about their missing documents and irregular data.

"Let me explain how this technology works," said Dhiren, a humanities lecturer in a government college just outside Guwahati city. He had been inducted as a disposing officer (DO) in the early months of 2019. The DO worked within a particular circle area and was the first—or last—human line of verification of the claims and objections that were filed by those who had failed their meetings with the FT. They reported to the Circle Registrar of Citizen Registration (CRCR, a circle officer in the administrative set up), who reported upward to the District Registrar of Citizen Registration (DRCR, a district commissioner in the administrative set up). Dhiren had been involved in the autonomy movement among the Karbi and Tiwa communities in the 1990s and early 2000s, before he got his job as a lecturer in 2013. He continues to be involved in cultural and social issues among indigenous communities, especially in the wider Kamrup, Morigaon, Nagaon, and Karbi Anglong areas. His work as the DO had kept him away from college, a fact that caused him some irritation. However, despite his personal misgivings about the nature of his work, he was all praise for the kind of technology that was being deployed in the NRC process. "Once the field-based work got over last year (in March 2018), things started getting a shape," he further explained about the family tree verification (FTV). As our conversation got into specifics, I had to keep track of the almost objective type, algorithm-based tenor of his descriptions.

"We had a domain for allowing for second and third generation respondents to make mistakes about the names of their immediate lineage relatives," he stated and added in the same breath: "'but can you forget the name of your own sister?" He claimed that many of the false claims of legacy data resulted from people giving different names for their immediate kin and siblings. He outlined the way in which the software was able to capture the inconsistencies in the manner in which certain persons claimed their family tree. Hence, lack of knowledge about immediate kin in an extended bilateral descent family would cause the software to determine that there was something amiss in the data provided. Dhiren was convinced that the software could not have got anything wrong, as far as catching on to the inconsistencies of personal narratives is concerned. Instead, as he explained the unfolding of a particularly poignant human drama, it was almost as if the software—in this case DOCSMEN—had begun to unravel family secrets into the public domain. When there was a mismatch in the family legacy (assigned an algorithm under the Legacy Data Code, or LDC), especially in the cases where two families claimed the same person and yet did not know anyone from the other family, they were asked to explain how the family trees for the same assigned LDC could go so wrong. In many cases, explanations attested to the frailty of human relationship: an aggrieved father who might have disowned a daughter for marrying against his wish, a man with a family in two towns that did not know one another, and so on. Others, Dhiren continued, were harder to let go. It was even more so when officials higher up the administrative chain had already verified the claims and objections at the investigative stage. In that case, the DO became the last human to deal with people who wanted answers from the executive body of the government.

The NRC process as it was rolled out in Assam relied on the convergence of technology, administrative efficiency, and political will in order to achieve its goals. However, as media reports show, there have been instances where Bengali-speaking persons have been subjected to unprecedented harassment for not being able to provide documents that could have made it through the NRC software (Mohan 2019). Often, as middle and lower level officials and part-time officers employed to conduct the NRC will attest, there has been pressure from political organizations and mid-level civil servants, including those in the courts of law and appellate bodies like the FT. In such cases, documents (and software) have the ability to strip away the context in which local interactions have become aggregated in the NRC process. They present a dilemma for students of social sciences who wish to research issues of citizenship. Twentieth-century citizenship research looked at the various ways through which an individual entered into a political and social relationship with the state that included economic welfare and civil rights (Marshall 1950), protection of cultural rights of minorities (Kymlicka 1996), and an overall protection of the sanctity of personal freedom that emanated from Kantian ideas of universally applicable rules that allowed individuals to be free. In a relatively short span of time since the Supreme Court instructed the government of Assam to conduct the NRC, one has been confronted by a different order of issues. For social scientists, then, the idea of generating data from a big, government and software-driven data gathering process is daunting. However, this is also the moment when one sees a greater need to put this data into a context. One needs to allow a plurality of narratives to emerge that challenges the current situation, where the state's narrative is considered superior to that of the neighbor, friend, and relative (Dourish and Cruz 2018).

Conclusion

Just as the dust was beginning to settle on the NRC debates, with every group sounding unhappy with the final list, the government of India introduced the Citizenship Amendment Bill of 2016 in parliament. The Bill became an Act of Parliament on December 11, 2019, resulting in immediate unrest in the Brahmaputra Valley of Assam and other states in the region. Urban and rural centers in the valley saw spontaneous uprisings and protests by Assamese-speaking people, as well others who were worried about the effect that the Act would have on politics in the region. The Act, detractors in Assam argued, had paved the way for undocumented Bengali Hindus to become citizens of India and settle legally. The protests spread to other parts of the country, where large sections of civil society argued that in the exclusion of Muslims, the government was attempting to create the legal basis for a majoritarian Hindu state. Even as the government made disingenuous attempts to placate the protestors, the unrest had brought about the halting of essential services and transport, resulting in the army being called back onto the streets of major cities to keep the peace. Tragically, six protestors were killed, and several prominent activists were arrested, all within a few days of the unrest in Assam.

It is not hard to see why my fellow traveler on the train was suspicious of a government-driven process where public health concerns were to be shared equally and equitably between the citizen and the state. In a way, his predicament was not very different from that of a villager without the right kind of documentation to prove their citizenship. Both were left at the mercy of the rules that were not very reassuring. Decades of structural violence had resulted in an ambivalent relationship marked cynicism and distrust of the state, as well as of one's neighbor. Subsequently, the language that has been employed to address the spread of the virus seems dangerously close to the kind of terminologies that were used during counterinsurgency. Color coded districts and words like "containment" have become commonplace in the vocabulary of those who have had to deal with a public health problem. It is probably fitting that his response was to evade the embrace of the state altogether, one that can hardly be successful in the present times, when the administration has taken it upon itself to isolate and segregate people as though it were still combatting insurgents among the population. Once again, the seemingly docile citizen (with or without documents) has become a cause for concern for the authorities, as there is a scramble to feed, detain, and quarantine people who lay claims to being from the region. This should serve as a moment to examine the impact that these debates will have on social science scholarship in the future. While it has disrupted relationships and forced people and organizations to reassess old colonial debates about autonomy and social justice, it has also highlighted the need to revisit new ones about transformation of the citizenship debates at a time when the world has been locked down by the rapid spread of a virus.

Barely three months after the anti-CAA protests began in Assam, the government of India's response to the spread of COVID 19 virus raised several questions about the nature of citizenship debates within the country. Despite the atomized existence that the lockdown measures have encouraged, it is difficult to ignore some of the optics that emerged in the course of the central government's response to pandemic. The plight of millions of migrants across India trying to make their way back to their home states and districts is one such reality that we have to confront. This migration was celebrated among certain circles as one of the outcomes of development. A decade on, it is clear for even the most dispassionate of observers that the treatment of migrants in the current lockdown is indicative of an underlying tendency for the central government to act as though internal migrants are less deserving of the rights bestowed upon citizens who are stuck overseas. Without recourse to trains, with uncertainties writ large about their access to food, the plight of internal migrants in India today echo that of the citizens of Assam during the brutal years of counterinsurgency: formally free, but always under the threat of extreme violence unleashed by a callous state. In an inchoate way these two entities—migrant and citizen—seem to be headed towards a shrunken little world of benign prisons and detention centers in the Indian subcontinent.

Note

1   See notification no. HCXXXVII-13/2017/2687/R.Cellof Gauhati High Court, June 21, 2017. https://districts.ecourts.gov.in/sites/default/files/Notification-21-06-2017_0.pdf.

References

Agamben, Giorgio. 2009. *What Is An Apparatus? and Other Essays,* translated by David Kishik and Stefan Pedatella. Stanford: Stanford University Press.

Azad, Abdul Kalam. 2018. "Growing up Miya in Assam: How the NRC weaponised my identity against me." *The Caravan,* September 23. https://caravanmagazine.in/politics/growing-up-miya-in-assam-how-the-nrc-weaponised-my-identity-against-me.

Barbora, Sanjay. 2016. "Introduction: Remembrance, Recounting and Resistance." In *Garrisoned Minds: Women and Armed Conflict in South Asia,* edited by Laxmi Murthy and Mitu Verma, 213–30. New Delhi: Speaking Tiger.

Barbora, Sanjay, and Saba Sharma. 2016. "Survivors of Ethnic Conflict." In *India Exclusion Report 2015,* edited by Harsh Mander, 206–228. New Delhi: Yoda Press.

Baruah, Sanjib. 2001. "Generals as Governors: The Parallel Political Systems of Northeast India." *Himal Southasian,* June, 10–20.

Baruah, Sanjib. 2019. *In the Name of the Nation: India and its Northeast.* Stanford: Stanford University Press.

Deka, Dixita. 2019. "Living without closure: memories of counter-insurgency and *secret killings* in Assam." *Asian Ethnicity*, July. https://doi.org/10.1080/14631369.2019.1639492.

Dourish, Paul, and Edgar Gómez Cruz. 2018. "Datafication and Data fiction: Narrating data and narrating with data." *Big Data & Society* 5, no. 2: 1–10.

Drohan, Brian. 2017. *Brutality in an Age of Human Rights: Activism and Counterinsurgency at the End of the British Empire.* Ithaca and London: Cornell University Press.

Kikon, Dolly. 2017. "Jackfruit seeds from Jharkhand: Being Adivasi in Assam." *Contributions to Indian Sociology* 51, no. 3: 313–37.

Kymlicka, Will. 1996. *Multicultural Citizenship: A Liberal Theory of Minority Rights.* Oxford: Clarendon Press.

Marshall, Thomas H. 1950. *Citizenship and Social Class and other essays.* London and Cambridge: The Syndics of the University Press.

Mohan, Rohini. 2019. "'Worse than a death sentence:' Inside Assam's sham trials that could strip millions of citizenship." *Scroll.in*, June 30. https://scroll.in/article/932134/worse-than-a-death-sentence-inside-assams-sham-trials-that-could-strip-millions-of-citizenship.

Rajkhowa, Gaurav, Ankur Tamuli Phukan, and Bidyut Sagar Boruah. 2018. "The Citizen Finds a Home: Identity Politics in Karbi Anglong." *Economic and Political Weekly* 53, no. 47: 17–20.

Roy, Anupama. 2019. "The Citizenship (Amendment) Bill, 2016 and the Aporia of Citizenship." *Economic and Political Weekly* 54, no. 49: 28–34.

Thompson, Robert. 1966. *Defeating Communist Insurgency: The Lessons from Malaya and Vietnam.* New York: Frederick A. Praeger.

Kalpana Kannabiran

Constitution-As-Commons:
Notes on Decolonizing Citizenship in India

This essay is dedicated to the memory of Adivasi leader Abhay Flavian Xaxa whose work has illuminated the elaboration of constitution-as-commons.

> **It will all be remembered (*sab yaad rakha jayega*)**
> . . . It will all be remembered
> yea each bit will be remembered
> while our hearts remain broken in the memory
> of our friends killed by your batons and guns
> we will remember each single thing, never forgetting
> knowing while you write your lies in ink
> but plainly in our blood perhaps
> the truth will surely be written . . .
> —excerpted from Aamir Aziz 2019[1]

The question of citizenship and the Constitution has been on the boil in India since 2014, when the Bharatiya Janata Party took over the reins of national government, the political turmoil finally erupting in 2019 simultaneously in several locales. There are stark continuities in the methods of rule deployed by the muscular, masculinist state in a neoliberal right wing context that privileges carcerality and militaristic control of peoples—"anti-nationals," "illegal infiltrators," "encroachers," and "untouchables." The prerogative to "'maim" (Puar 2017) has proliferated in the past five years, spreading from state actors to virtual and physical mobs self-identifying with the ruling Hindutva regime. The flip side of the politics of maiming is docility that congeals as the most desirable institutional attribute, especially in higher education,

The South Atlantic Quarterly 120:1, January 2021
DOI 10.1215/00382876-8795890 © 2021 Duke University Press

but also in the justice system. Bigotry emerges as the new, legitimate normal. This template, with vigilantism as the method, powered by lynch mobs armed with viral messaging platforms, is deployed in multiple contexts to desired effect. The politics of maiming fuels technologies of dispossession, colonization, occupation, and incitement to hate crime in the present juncture, under the watch of the Constitution and special statutory protections for communities vulnerable to discrimination and structural violence.

The rule by maiming injects discourses of contagion and methods of vigilantism honed to precision in violently exclusionary fields of caste, bigotry, hate, and terror into a pandemic context. Languages of war against "enemies of the state" morph into the call to arms against a virus in a volatile, Islamophobic social context seamlessly enfolding viralities of contagion within viralities of religious profiling and untouchability practices. The Constitution is silenced through practices of juridicalization that reinstate responsibilization as the sign of citizenship (Tella 2020).

I revisit my earlier arguments on nondiscrimination and liberty (Kannabiran 2012) at a time when intensely vindictive carcerality has become the defining trait of the state and equivocation/jurisprudential dissociation in matters of life and personal liberty has come to represent juridicalization *in the name of* the Constitution.

The collective Dalit-Muslim-Adivasi-Kashmiri experience of annihilation, occupation, carcerality, and crimes against humanity—and the multiplicity of their resistance to the rule by maiming, as also the archives of resistance they draw upon—provides the substance for the cursory exploration of the idea of constitution-as-commons and its ethnographies that follows. This is an insurgency to wrest citizenship from the jaws of enclosure, through a radical crafting of the contours of the constitution-as-commons. The following sections will provide a sketch of the constitution-as-commons and the resistance against its enclosure.

Signposting the Constitution-as-Commons

The well is the landlords (*kuan thakur ka*)
the oven is made of mud
the mud is from the lake
the lake is the landlords.
we hunger for bread
bread made from millets
the millets are from the field
the field is the landlords.

the bullock is the landlords
and the plough
but the hands on the plough are ours.
the harvest is the landlords
the well is the landlords
the water too
the crops and fields are his
the lanes and streets are his.
then what, pray, are ours?
the village?
the city?
the country?
Omprakash Valmiki (1950–2013), ?1980

The idea of the constitution-as-commons resurrects the moral economy of the constitution and its ethical foundations immersed in empathy ("fraternity"), most evident in the work of B. R. Ambedkar. In a situation of extreme repression and the near-total capitulation of constitutional courts on Article 21 (the right to life and personal liberty), the unfolding of a shared imagination for the present-future, the resurgence of collective action in hitherto unknown ways, and the crafting of social capital around the radical public reading of the Constitution (notably the preamble) defines constitutional communities in deeply insurgent ways. There is a shift in legitimacy from the formal holders of constitutional power to the multitudes that deliberate on the Constitution (on streets and campuses) and explore/interpret its meanings. The constitution-as-commons posits the Constitution (the text) as the irreducible ethical framework that must bind legislatures, courts, and governments alike, asserts the indivisibility of the Constitution, and claims free and unfettered access to the Constitution and its protections. In this scheme, the preamble (in letter and spirit) as the anthem of resistance, anchors the understanding of the Constitution.

Over the past couple of years (at least), we have been witness to recitations of *lyrical constitutionalism* in constitutional jurisprudence (Kannabiran 2019). There are older and more enduring traditions that predate the Constitution and anticipate a radical, just, insurgent constitutionalism as the ethical basis for independent India—the songs of Ambedkari Shahirs, or Ambedkari poets are an example (Maitreya 2018). We also have the stunning delineation of the constitution-as-commons by Dalit poet Omprakash Valmiki above, anticipating the performances of the present time.

The Constitution may be crafted around conversations on the commons, insurgent politics, and democracy. One possible way to do this could be through an interreading of B. R. Ambedkar and Ambedkarite writing and performance and the vast traditions of resistance on the Indian subcontinent on the one hand, and the work of Elinor Ostrom and the commons scholars, Antonio Negri's and Michael Hardt's vast corpus on constituent power and resistance, Gloria Anzaldua's (1987) work on borderlands, Walter Mignolo and Catherine Walsh's (2018) work on *"decolonial pluriversality* and *pluriversal decoloniality,"* and Martha Jones' (2018) recent work on *"birthright citizenship,"* on the other, for instance. While a detailed exploration is outside the scope of the present essay, my specific concern here is with exploring the Indian Constitution as vehicle and medium of resistance and retrieval, taking forward the work of Upendra Baxi and K. G. Kannabiran. Central to this endeavor is the redefinition of "constitutional communities" in a manner that engages with popular constitutionalism, and the meanings this collective wisdom brings anew to the constitution-as-commons.

The Constitution stands in opposition to the contracommons (not the anticommons)—dismantling through public action, the structural aporias that fuel the enclosure of the Constitution, indeed its capture. How may we discover the enclosure threats posed by the contracommons and the travesties it engenders? How have contracommons regimes been constituted, and how may we trace the source of their Power? In an inversion, we could perhaps plot the tragicomedies of the contracommons that provide glimpses of hope and recovery to the constitution-as-commons, as for instance the resurrection of dissents in the celebrated decision upholding the right to privacy as a fundamental right under the Constitution in 2017.[2] This decision alone helps a calibrated reading of the constitution-in-courts, disentangling the articulation of the constitution-as-commons from the contracommons therein. This is particularly relevant as we witness the emergence of a supremacist authoritarian regime that collapses legislative action, state repression, and judicial interpretation together, so that the first important distinction in modalities of action/acting is between *occupying* the constitution-as-commons (spatially and ideationally) and the juridicalization in the contracommons.

Resisting the Travesties of the Contracommons

Rohith Vemula (1989–2016) led the resistance against the siege of higher education by majoritarian government and the cascading movement of Dalit students around social justice and Ambedkarite philosophies on university campuses, especially after 2014. The quelling of resistance through criminalization and the securitized enclosure of campuses mark this moment. At the present time, the Bhima Koregaon case that saw a spate of arrests of political dissenters, human rights defenders, and Dalit cultural activists and intellectuals, signals the now standard state response to Dalit assertion, and in a twist recognizes the enduring ways in which Dalit resistance shapes the politics of constitutional insurgency in India generally. The arrest of the widely respected Dalit scholar Anand Teltumbde on April 14, 2020, and the arrest of Dalit leader Chandrashekhar Azad, founder of the Bhim Army a few months earlier, are illustrative. Azad's reading of the preamble to the Constitution on the steps of one of the oldest mosques in Delhi was construed as inflammatory and an incitement to violence (Singh 2020).

The sounding of the bugle of anti-caste protest and Dalit assertion of the constitution-as-commons invites the most repressive state action enforced through its juridicalization. The targeted violence against Muslims in Delhi in February 2020, in the state of Uttar Pradesh in 2019 (Citizens Against Hate 2020), and the ongoing incarceration of dissenting Muslim students, youth, and journalists follows the widespread protests against the *Citizenship Amendment Act*, 2019, which introduced a religious basis for citizenship claims, specifically excluding Muslims. Just prior to this was the decision of the Supreme Court of India in the matter of Babri Masjid that handed over the site of the demolished mosque to Hindus who claimed to represent the deity. It must be recalled that the demolition of the mosque by Hindu mobs in 1992 triggered one of the worst episodes of mass violence against Muslims in the country. In challenging a law that undermines the fundamental basis of the Constitution of India, Muslims have led the resistance against the introduction of the denominational basis for citizenship. Poet Aamir Aziz's poignant poem (2019) *"sab yaad rakha jaayega"* (It will all be remembered), quoted in the epigraph, resonates through the protests and writing on citizenship and resistance.

There is a two-pronged strategy adopted by the majoritarian state to set this law of denominational dispossession in motion—the first is through a concerted armed assault on protestors (mostly women)—protests in the Shaheen Bagh (Shaheen Bagh Official 2020) suburb of Delhi are illustrative of this larger struggle; the second is through the mass incarcerations of poor people in the state of Assam, mostly Muslims who live on the edge of precarity now excluded from the National Register of Citizens and labelled *ghoospethiye* (a stigmatizing word that means "infiltrators"). The popular Muslim resistance against the Citizenship Amendment Act, 2019 recovers the constitution-as-commons by rejecting the use of religious faith as a ground of discrimination.

Article 370 of the Constitution of India guaranteed autonomy for the state of Jammu and Kashmir, and Article 35A protected the state from transfer of lands to non-residents. Both these provisions together were part of a constitutional compact of autonomy and self-determination for Kashmir. On August 5, 2019, the Indian parliament abrogated Article 370 and 35A of the Constitution of India and split of the state into two Union Territories—Jammu and Kashmir and Ladakh. This erasure of statehood has a long and troubled history of military occupation, struggles for self-determination, militancy, the suspension of the rule of law, grief, loss and everyday resistance, and tenacious noncooperation, especially by women and young people. Demands for justice and peace in Kashmir, and for the return of life untroubled by militarization, securitization, and its dark perils on an everyday level, have been silenced through state violence and impunity to the armed forces. The report of an all-women team that visited the valley in February 2020 details the spiraling effect of the post-abrogation lockdown on everyday life and socialities in Kashmir—arbitrary arrests and detentions of Kashmiri youth, aggravated surveillance, the suspension of media freedoms, the suspension of internet in the valley, the loss of jobs and incomes, sale of land and assets to meet living expenses and medical emergencies, the blocking of all routes to decent work for fair wages—the majority pushed to the edge of precarity by the state (Kannabiran et al. 2020). The call for *azaadi* (freedom) that we hear from Kashmir is a reinstatement of the constitution-as-commons—one that respects and guarantees the right to autonomy, dignity, and self-determination.

The final piece in this account of the contracommons speaks of adivasis/tribes in India, protected both by the nondiscrimination provision of the constitution, but also importantly by the guarantee of territorial autonomy and self-governance to tribal homelands under the fifth and sixth schedules of the Constitution. The figure of Abhay Xaxa, the forty-year-old Adivasi leader from Chhattisgarh who died unexpectedly on March 14, 2020, tells the story of Adivasi resistance and forest perspectives on the constitution-as-commons—"indigenocracy" as he called it. In 2015, the Modi government attempted to push through a bill to amend the Right to Fair Compensation and Transparency in Land Acquisition, Rehabilitation and Resettlement Act, 2013, a move that would have divested Adivasi communities of land and common resources. In a resistance led by Xaxa, around sixty Adivasi men gathered in front of a local government office and publicly defecated on copies of the proposed bill. In a response to the widespread reaction to this stunning protest, Xaxa asserted, "Poop protest is the most peaceful and democratic protest against a black law which threatens the core of their life- Jal (water), Jangal (forest) aur Jameen (land), that is! Poop protest is not a new form of agitation, In fact throughout history, whenever oppressed masses have dropped their shit as arsenal, rulers have been shaken because it often marks the beginning of a social uprising."

Despite a sustained focus by social movements, and the convergence of insurgent law-making and critical ethnography, it has still been possible for the Supreme Court in 2018 to follow the trail of wildlife conservationists in labelling indigenous forest dwellers "encroachers" and directing their eviction from forests, rolling the Constitution back through juridical moves. Stopped in its tracks by widespread protests from Adivasi and forest-dwelling communities across the country, the Supreme Court stayed the eviction during the pendency of the hearing (still ongoing). Xaxa attributed this juridical volte-face to the fact that "the supreme court was told that 2 million Adivasis are sharpening their bows and arrows before marching to Delhi . . . No benevolence" (Choudhury and Aga 2020).[3] The deep entrenchment of dominant majoritarian claims in juridical habits that reinforce dominion over Adivasis—and "epistemicide"—surfaced yet again (this time in the middle of the COVID 19 lockdown, which rendered mass protest impossible) when, on April 22, 2020 the Supreme Court reversed the government order in the southern state of Andhra Pradesh, which mandated that only Adivasi school teachers would be appointed in publicly funded schools in Adivasi homelands (protected under Schedule 5 of the Constitution). Rolling back special protections, the court observed that they were put in place for tribes in the first instance because "their language and *their primitive way of life makes them unfit* to put up with the mainstream and to be governed by the ordinary laws."[4] This systematic distortion of the historical archive and of history fence in the constitution in each instance cited in this section.

Abhay Xaxa's understanding that the Supreme Court's juridicalization, and thereby its negation of the right of Adivasis to the constitution-as-commons, is based on perspectives from Brahmanical environmentalism is sharp and precise, mirroring in a sense the assertion of Winona LaDuke and Deborah Cowen (2020: 257): "We want to be at the table, not on the menu."

Enclosure that shrouds the country remains securitized and deathly in the pandemic context—we see the refusal of the state to lift restrictions on internet in Kashmir despite its own COVID-19 advisory, and the auctioning of mines in Kashmir while the double lockdown is in place; we see the brazen profiling of Muslims as the carriers of contagion by state and media alike; we see the request of the National Investigation Agency to use handcuffs on

Anand Teltumbde, claiming the need to avoid physical contact in a viral context; we bear painful witness to the death from exhaustion of Jamalo Madakam, a twelve-year-old Adivasi girl, after walking 150 kilometers to her hometown during lockdown, from the chilly farms where she was put to work (her figure standing in for the exodus of displaced workers and their countless deaths from the lockdown—not from COVID 19 infection). Viralities, suffering and gruesome death are selective of social location. The dismantling of the contracommons is in fact a matter of life, death, and freedom.

Clearing a Path to the Present-Future

Even while we witness a consolidation of ruthless state power and the normalization of impunity via the contracommons, the borderlands come to symbolize the nation in the throes of disembowelment. The preamble to the Indian Constitution as the anthem of the anti-CAA protests soared to a crescendo at protest sites across the country in 2019. The return to songs of freedom and resistance against a vindictively carceral state—the Kashmiri Bella Ciao (Wanaan 2020), the multi-lingual performance of Urdu poet Faiz Ahmed Faiz's poem *Hum Dekhenge* (We shall see) at protests against the CAA across the country, and the birth of a new genre of poetry—Miya poetry by Muslim poets in Assam that turned a stigmatizing label into self-assertion (Daniyal 2019)—present to us the visceral, enduring insurgent possibilities of contemporary resistance. For this carries echoes of generations of insurgencies that asserted "birthright citizenship." Insurgent readings and B. R. Ambedkar's sharp critique of Hindu religion—the inseparability of the annihilation of caste from the dismantling of the Hindu social order (its exclusions, segregations, violence, oppressions), and his infinite corpus on life and politics are constitutive of this ongoing insurgency. This moment presents to us an opportunity to map the fields of constitutionalism anew, eschewing languages of war and enemies (along with its attendant proxy fences and securitized borders), inscribing instead decolonial visions of the constitution-as-commons—dignity, autonomy, self-determination, birthright, and justice—drawing on a history of the present from the perspectives of the cascading resistance in the borderlands.

Notes

1   The translation of the poems by Aamir Aziz (Urdu) and Omprakash Valmiki (Hindi) by Vasanth Kannabiran, are published in this essay for the first time. I am grateful to her for doing this "on demand." I am grateful to Aamir Aziz for permission to quote this translated excerpt.

2   I borrow the term 'tragicomedy' from Hess and Ostrom 2007.In *Puttaswamy v. Union of India* (AIR 2017 SC 4161) a nine-judge bench of the Supreme Court of India unanimously declared the right to privacy a fundamental right under the constitution, and reinstated 3 dissenting opinions between 1950 and 1975 that upheld the right against state surveillance, the indivisibility of fundamental rights and their non-derogability even in conditions of emergency.

3   All references to Abhay Xaxa are from this article.

4   *Chebrolu Leela Prasad Rao and Others vs State of Andhra Pradesh and Ors.* Civil Appeal No. 3609/2002, dated April 22, 2020. MANU/SCOR/24647/2020.

References

Anzaldua, Gloria. 1987. *Borderlands/La Frontiera: The New Mestiza.* San Francisco: Aunt Lute Books.

Choudhury, Chitrangada, and Aniket Aga. 2020. "In Memoriam: Sociologist and Activist Abhay Xaxa." *India Forum*, April 3. theindiaforum.in/article/memoriam-sociologist-activist-abhay-xaxa.

Citizens Against Hate. 2020. *Everyone Has Been Silenced: Police excess against anti-CAA protesters in Uttar Pradesh, and post-violence reprisal.* New Delhi: Citizens Against Hate. https://www.hrfn.org/wp-content/uploads/2020/03/Everyone-Has-Been-Silenced.pdf.

Daniyal, Shoaib. 2019. "'I Am Miya': Why Poetry by Bengal-Origin Muslims in Their Mother Tongue Is Shaking Up Assam." *Scroll.in*, June 14. scroll.in/article/930416/i-am-miya-why-poetry-by-bengal-origin-muslims-in-their-mother-tongue-is-shaking-up-assam.

Hess, Charlotte, and Elinor Ostrom. 2007. "Introduction: An Overview of the Knowledge Commons." In *Understanding Knowledge as a Commons: From Theory to Practice*, edited by Charlotte Hess and Elinor Ostrom, 10. Cambridge, MA and London: MIT Press.

Jones, Martha S. 2018. *Birthright Citizens: A History of Race and Rights in Antebellum America.* Cambridge: Cambridge University Press.

Kannabiran, Kalpana. 2012. *Tools of Justice: Non-Discrimination and the Indian Constitution.* New Delhi: Routledge.

Kannabiran, Kalpana. 2019. "What Use is Poetry: Excavating Tongues of Justice around Navtej Singh Johar v. Union of India." *National Law School of India Review* 31, no. 1, 1 –31.

Kannabiran, Kalpana, Sarojini Nadimpally Navsharan Singh, Roshmi Goswami, and Pamela Philipose. 2020. "Interrogating the "Normal" in Kashmir: Report of a Visit to the Valley, January 31 to February 5, 2020." *Indian Cultural Forum*, March 4. indianculturalforum.in/2020/03/04/interrogating-the-normal-in-kashmir/.

LaDuke, Winona, and Deborah Cowen. 2020. "Beyond Wiindigo Infrastructure." *South Atlantic Quarterly* 119, no. 2: 257.

Maitreya, Yogesh. 2018. "In the Verses of Dalit Shahirs, You Can Hear the History of India's Anti-Caste movement." *Scroll.in*, June 9. https://scroll.in/magazine/878456/in-the-verses-of-dalit-shahirs-you-can-hear-the-history-of-indias-anti-caste-movement.

Mignolo, Walter D., and Catherine E. Walsh. 2018. *On Decoloniality: Concepts, Analytics, Praxis.* Durham, NC: Duke University Press.

Puar, Jasbir K. 2017. *The Right to Maim: Debility, Capacity, Disability.* Durham, NC: Duke University Press.

Shaheen Bagh Official (@Shaheenbaghoff1). 2020. *Twitter*. twitter.com/shaheenbaghoff
1?lang=en.

Singh, Aditi. 2020. "Delhi Court Grants Bail to Bhim Army Chief Chandra Shekhar Azad
(Subject to Several Conditions)." *Bar and Bench*, January 15. barandbench.com/news
/litigation/breaking-delhi-court-grants-bail-to-bhim-army-chief-chandra-shekhar-azad
-subject-to-several-conditions.

Tella, Ramya K. 2020. "Responsibility in Viral Times: A Note from India." *Discover Society*,
March 30. https://discoversociety.org/2020/03/30/responsibility-in-viral-times-a
-note-from-india/.

Wanaan, Zanaan. 2020. "Kashmiri Bella Ciao." *YouTube*, February 23. youtube.com/watch
?v=ABe4V1DRojM.

Notes on Contributors

Joel Auerbach is a doctoral student in rhetoric at the University of California, Berkeley. He has an MA from McGill University and a BA from Vassar College.

Nandita Badami is a doctoral candidate in anthropology at the University of California, Irvine. Her previous degrees include an MA and an MPhil in political studies from Jawaharlal Nehru University, New Delhi.

Daniel A. Barber is an associate professor at the University of Pennsylvania Weitzman School of Design, where he chairs the PhD program in architecture. His most recent book is *Modern Architecture and Climate: Design before Air Conditioning*. He codirects the Penn/Mellon Seminar on Humanities, Urbanism, and Design and coedits the *Accumulation* series on e-flux architecture.

Sanjay Barbora is a sociologist and civil rights activist, and Associate Professor at the Tata Institute of Social Sciences, Guwahati.

Darin Barney is the Grierson Chair in Communication Studies at McGill University. He is author of several books and articles on infrastructure, technology, and politics including, most recently, the coedited volume, *The Participatory Condition in the Digital Age* (2016).

Bharat Bhushan is a journalist based in Delhi.

Amanda Boetzkes is Professor of Contemporary Art History and Theory at the University of Guelph. She is the author of *Plastic Capitalism: Contemporary Art and the Drive to Waste* (2019) and *The Ethics of Earth Art* (2010), and coeditor of *Heidegger and the Work of Art History* (2014).

Dominic Boyer is Professor of Anthropology at Rice University, Founding Director of the Center Energy and Environmental Research in the Human Sciences (CENHS), and author of *The Life Informatic: Newsmaking in the Digital Era* (2013) and *Energopolitics: Wind and Power in the Anthropocene* (2019).

Partha Chatterjee is Professor of Anthropology at Columbia University.

Sabyasachi Basu Ray Chaudhury is Professor, Department of Political Science, Rabindra Bharati University, Kolkata, India and Member, Calcutta Research Group.

Jamie Cross is Professor of Social and Economic Anthropology at the University of Edinburgh. He is the author of *Dream Zones: Anticipating Capitalism and Development in India* (2014).

Gökçe Günel is Assistant Professor in Anthropology at Rice University. Her latest book is *Spaceship in the Desert: Energy, Climate Change, and Urban Design in Abu Dhabi* (2019).

Eva-Lynn Jagoe is an associate professor of comparative literature and Latin American studies at the University of Toronto. Her most recent book, *Take Her, She's Yours* (2020), explores theoretical notions of psychoanalysis, subjectivity and feminism through an experiential first-person narrative.

Kalpana Kannabiran is Professor of Sociology and Director, Council for Social Development, Hyderabad. She is a member of the Calcutta Research Group.

Jordan B. Kinder is Research Director of the Petrocultures Research Group and incoming SSHRC Postdoctoral Fellow at McGill University. His work can be found in *Energy Culture* (2019), *The Bloomsbury Companion to Marx* (2018), *Mediations, Socialism and Democracy*, and elsewhere. He is a citizen of the Métis Nation of Alberta.

Ranabir Samaddar is the Distinguished Chair in Migration and Forced Migration Studies at the Calcutta Research Group and belongs to the school of critical thinking.

Mark Simpson is Professor in English and Film Studies at the University of Alberta (Treaty Six / Métis territory). His recent contributions on aspects of petroculture include "Lubricity" (*Petrocultures* 2017), "Kerosene" (*Fueling Culture* 2017), "Five Theses on Sabotage in the Shadow of Fossil Capital" (with Jeff Diamanti; *Radical History* 2018), and "Resource Aesthetics" (with Brent Bellamy and Michael O'Driscoll; *Postmodern Culture* 2016).

Nicole Starosielski is Associate Professor of Media, Culture, and Communication at New York University. She is author of *The Undersea Network* (2015) and coeditor of *Signal Traffic: Critical Studies of Media Infrastructure* (2015), *Sustainable Media: Critical Approaches to Media and Environment* (2016), and *Assembly Codes: The Logistics of Media* (forthcoming).

Imre Szeman is University Research Chair in Communication Arts at the University of Waterloo. His recent work includes *On Petrocultures: Globalization, Culture, and Energy* (2019) and *Energy Culture: Art and Theory on Oil and beyond* (coedited with Jeff Diamanti, 2019).

Rhys Williams is Lecturer in Energy and Environmental Humanities at the University of Glasgow.

Sheena Wilson is Professor of Media, Communications, and Cultural Studies at the University of Alberta, on Treaty No. 6 territory. Publication highlights include "Energy Imaginaries: Feminist and Decolonial Futures" (2018); *Petrocultures: Oil, Politics, Cultures* (with Adam Carlson and Imre Szeman, 2017); and "Gendering Oil: Tracing Western Petrosexual Relations" (2014).

DOI 10.1215/00382876-8912469